S.F Ergashev
A.A Kuchkarov

DOMESTIC SOLAR WATER HEATERS

S.F Ergashev
A.A Kuchkarov

DOMESTIC SOLAR WATER HEATERS

ScienciaScripts

Imprint

Any brand names and product names mentioned in this book are subject to trademark, brand or patent protection and are trademarks or registered trademarks of their respective holders. The use of brand names, product names, common names, trade names, product descriptions etc. even without a particular marking in this work is in no way to be construed to mean that such names may be regarded as unrestricted in respect of trademark and brand protection legislation and could thus be used by anyone.

Cover image: www.ingimage.com

This book is a translation from the original published under ISBN 978-620-4-72589-5.

Publisher:
Sciencia Scripts
is a trademark of
Dodo Books Indian Ocean Ltd. and OmniScriptum S.R.L publishing group

120 High Road, East Finchley, London, N2 9ED, United Kingdom
Str. Armeneasca 28/1, office 1, Chisinau MD-2012, Republic of Moldova, Europe
Printed at: see last page
ISBN: 978-620-5-27139-1

Contents

In recent years, the use of solar energy in industry, utilities, agriculture and households has been expanding in many countries around the world. Solar energy cannot fully supply the energy needs of an industrial, municipal, agricultural or domestic facility, but it can save a significant portion of traditional energy resources. Solar energy is therefore seen as an additional source of energy.

Solar water heating systems are widespread in different countries of the world, such as Canada, USA, Germany, China, India and others. They are also of great interest in the CIS, in particular in Russia and Ukraine, the Central Asian republics: Uzbekistan, Turkmenistan, Kyrgyzstan, Kazakhstan. Domestic solar water heating plants are designed to produce hot water in spring and summer by converting the energy of solar radiation into thermal energy, without using traditional sources - electricity, solid fuels or liquid fuels. Their widespread use gives a significant saving of fuel and energy resources.

The tasks set in the monograph are a continuation of the work carried out by the authors together with the ENIN named after G. Krzhizhanovsky "Development, research and setting up industrial production of simple and inexpensive designs of household solar water heaters for individual consumers". Based on the results of the above programme, various prototypes of solar water heating systems were developed and created. In particular, the thermo-technical characteristics and performance of solar household water heating plants, depending on climatic conditions and the intensity of solar radiation, were investigated.

The experience gained by the authors in developing and testing various versions of solar domestic water heating systems has been used in fulfilling the objective of this project, i.e. in setting up industrial production of a simple and cheap design of a domestic conical-type solar water heater at affordable prices for individual consumers.

CHAPTER I
ANALYSIS OF THE CURRENT STATE OF RESEARCH AND DEVELOPMENT OF SOLAR DOMESTIC WATER HEATERS

At present, the most researched and prepared for widespread practical application are domestic solar water heaters designed for individual consumers. Sola r domestic water heaters are a modern and efficient system for producing hot water water by absorbing solar radiation, converting it into heat, accumulating it and transmitting it to the consumer.

Solar domestic water heaters have become the ideal solution for homes, flats, swimming pools, hotels, hospitals, laundries and various industrial sectors where fast and efficient water heating is required with overall fuel savings.

History. The first solar water heater was created in 1767 by a Swiss botanist **Oras Benedict de Saussure** and in terms of its capacity, it allowed for the cooking of **soup** [1].

The modern type of water heater was developed in 1953 by Israel by engineer Levi Issar, and perfected by Dr. Zvi Tavor in 1955, for which he received a prize of 1,000 Israeli liras from the Prime Minister 3 years later, David Ben-Gurion [1].

§1.1 Design of the solar water heater

The specificity of modern solar water heaters lies in the fact that their solar cells are made of special materials that maximise heat absorption due to their high sensitivity to the thermal spectrum of the sun's rays.

Generally speaking, solar heaters are quite simple in design, consisting of three main parts:

- A solar flat or vacuum collector, which is responsible for absorbing the sun's energy, ensuring it is transferred to the water;
- a thermal storage tank (heat exchanger tank);
- connecting pipes.

Fig.1.1 Structure of the solar water heater [2].

The structure of a vacuum solar water heater consists of four elements:

- hot water storage tank
- stainless steel case
- reflectors for dual use of the sun's rays
- vacuum tubes made of borosilicate glass that absorb the sun's rays and transfer heat to the water

Figure 1.2. borosilicate glass vacuum tubes [2].

The heater includes a magnesium anode rod to protect against corrosion and accumulate mineral salts in the case of excessively hard water. This method reduces excessive concentrations of elements such as calcium, magnesium, lithium, zinc, etc. in the water.

The glass vacuum tubes are fitted with a special protector layer and have high impact resistance, with a safety margin capable of withstanding even 2 centimetre-high hailstones.

The systems are installed on the roof of the house, and are oriented properly towards the sun, i.e. according to the most convenient position relative to the sun at an angle to the horizon, with the orientation towards the south.

The water circulates through the system due to the so-called thermosyphon effect, which provides the required temperature difference.

§1.2 Flat solar collector.

The flat-plate collector is the most common type of solar collector used in domestic water heating and heating systems. This collector is a thermally insulated glazed panel in which an absorber plate is placed. The absorber plate is made of a metal that conducts heat well (usually copper or aluminium). Copper is used most often, as it conducts heat better and is less prone to corrosion than aluminium. The absorber plate is treated with a special highly selective coating that better retains the absorbed sunlight. This coating consists of a very strong thin layer of amorphous semiconductor deposited on a metal substrate and is highly absorptive in the visible spectrum and low in the longwave infra-red range. Thanks to the glazing (flat-plate collectors usually use frosted, light-only, low-ferrous glass), heat losses are reduced. The bottom and side walls of the collector are covered with insulating material, which further reduces heat loss.

The principle of operation of a flat-plate collector. Sunlight passes through the glazing and hits the absorber plate, which heats up, converting the solar radiation into thermal energy. This heat is transferred to the heat transfer medium - water or antifreeze - circulating through the solar collector. The heat transfer medium heats up and then transfers the heat energy via a heat exchanger to the water in the cylinder. The hot water remains there until it is used. An electric insert can also be installed in the cylinder so that if the temperature drops below the set temperature (e.g. due to prolonged cloudy weather), it heats the water up to the set temperature.

Figure 1.3 Diagram of a flat-plate collector [2].

Vacuum flat-plate collector. In the vacuum flat-plate collector, the volume containing the dark surface that absorbs solar radiation is separated from the surrounding area by an evacuated space, which virtually eliminates heat loss to the environment through heat conduction and convection. Radiation losses are largely suppressed by the use of a selective coating. As the total loss factor in a vacuum collector is small, the heat transfer medium can be heated there to temperatures of 120 - 160°C .The solar vacuum collector ensures the collection of solar radiation in all weathers, almost regardless of the outside temperature. The energy absorption coefficient of such collectors, at vacuum levels of 10-5 and 10-6, is 98 %. Insulation in the form of a vacuum avoids heat loss [3].

Figure 1.4: Vacuum flat-plate collector [3].

§1.3 Vacuum solar collectors with glass vacuum tubes.

They consist of heat tubes, which resemble a thermos in structure, but with a transparent outer part of the tube. The inside of the tube is coated with a special substance that absorbs sunlight. The tubes can also act as a heat conductor. Such tubes contain water at the bottom, which is heated and turned into steam. Vacuum collectors can heat water up to 250-300 degrees (with limited heat extraction), and up to boiling temperatures at negative ambient temperatures [3].

Figure 1.5: Vacuum tubes [3].

Through the use of heat pipes in the design of vacuum collectors, greater efficiency is achieved when operating at low temperatures and low light conditions. At the same time, the use of an additional heat loop leads to unavoidable losses associated with heat transfer between media, so at temperatures above +15 degrees the efficiency of vacuum collectors is almost the same, and sometimes even lower, than that of flat-plate collectors. Due to the high-quality multi-layer highly selective coatings and vacuumisation, a modern solar collector is capable of capturing solar energy in a very wide spectrum of radiation (considerably wider than the visible spectrum).

There are several basic types of vacuum solar collectors:

1. Flask within a flask.

2. Flask in flask with heat pipe.

3. Vacuum flask.

Flask within a flask. In the first type of collector, the heat transfer medium is heated by contact with the selective coating of the glass bulb. The heat transfer medium can be either water or antifreeze (or a mixture of water and antifreeze). Such systems operate without overpressure on the coolant side as they cannot be effectively waterproofed. These systems are most often passive circulation systems.

Bulb in bulb with heat pipe. Collectors using the second type of flask use copper heat pipes. Heat transfer from the absorber to the tube is carried out by means of fins. The heat pipe transfers the heat to the heat pipe condenser, which is connected to the collector where the heat transfer medium is circulated.

Vacuum flask. The main difference of the third type of flask is the vacuum heat insulation of the copper heat pipe. Whereas in flasks of the first and second type the vacuum layer is located between the glass walls of the flasks, in vacuum-insulated flasks both the absorber and the heat pipe are at reduced air pressure. In addition, only one layer of glass instead of two increases the **EFFICIENCY** installations.

Advantages and applications of vacuum solar collectors. Due to their high thermal insulation, vacuum solar collectors operate very efficiently at low ambient temperatures. The advantage of vacuum collectors begins to show at air temperatures below -15 degrees Celsius. At negative air temperatures there is no alternative to vacuum collectors. In the photo taken with a thermal imaging camera you can see the difference in heat loss between the vacuum collector (left) and the flat-plate collector (right) [4].

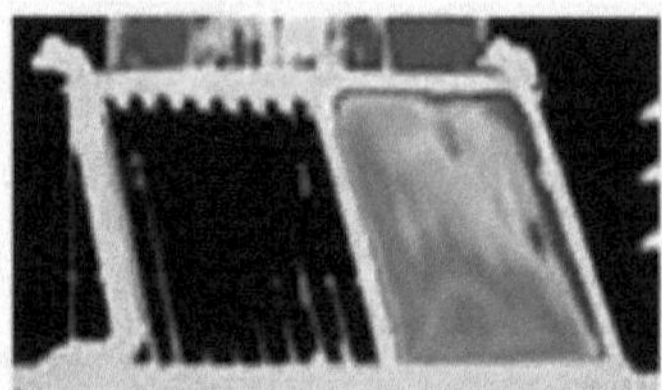

Figure 1.6. Difference in heat loss between vacuum collector (left) and flat-plate collector (right). [4].

Solar thermal systems based on vacuum collectors can be used for both hot water and home heating purposes. In this case, it is possible to obtain hot water entirely from the solar heater in the summertime. During the rest of the year, up to 60 % of the hot water can be obtained from solar energy.

A solar vacuum collector collects solar radiation in all weathers, reducing dependence on outside temperature. The energy absorption coefficient of the collectors reaches 98%, but due to the losses associated with the reflection of light by the glass tubes and their incomplete light permeability, it is lower.

The efficiency of solar collectors can, as a first approximation, be calculated using the following formula:

$$\eta = \eta_0 - \frac{k \cdot \Delta T}{E},$$

where η - calculated efficiency value, - nominal (optical) efficiency of the unit under normal conditions, k_1 - coefficient depending on the type and thermal insulation of the collector, - temperature difference between the heat transfer medium and the ambient air (gr C), E - insolation (W/sq.m.).

Data for some types of collectors are given below.

Collector type	Nominal efficiency	Factor k_1
Flat solar collector	72-75	3-5
Vacuum solar collector with heat pipes	60-65	0,7-1,1
Plastic solar collector	50-60	Up to 80

Solar thermal water heaters. 2009, the capacity of existing solar water heating systems increased by 21%, reaching a level of about 180 GW of thermal capacity. China installed 29 GW, or about 42 million m^{3^2} . Most of the remaining capacity is concentrated in the European Union, where 2.9 GWh were installed in 2009 (about 4 million metres^2). Of the European Union countries, Germany has made the most progress (1.1 GWh or 1.6 million m^2) [5].

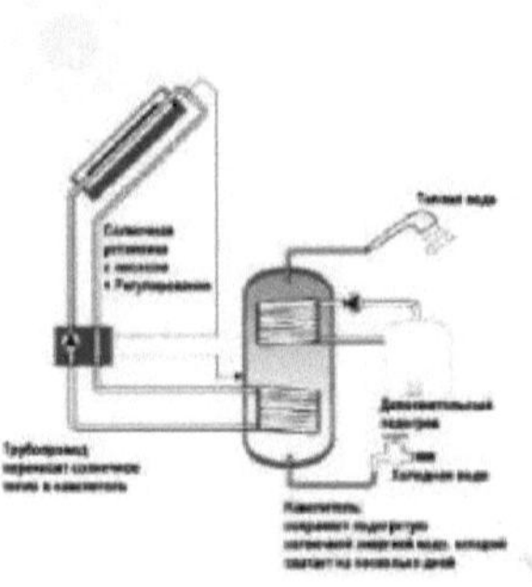

Figure 1.7: Operating principle of a solar water heating system [5].

The simplest and cheapest type of water heater is a collector without a glass cover. This type of water heater consists of a black absorber, with no translucent coating, thermal insulation or housing. Such collectors are often made of synthetic materials and can be used for heating water in swimming pools during the summer.

The water circulation in the closed solar collector-tank-solar collector circuit can be forced by a small circulation pump or naturally by the hydrostatic pressure differential between the cold water column and the heated water column. In the latter case, the tank must be positioned sufficiently high, preferably above the top of the solar collector.

Figure 1.8. Flat collector water heater developed by Belarusian State University of Informatics and Radioelectronics". [5].

Solar water heating systems (SHS) have been developed and successfully implemented in Belarus since 2000. They are intended for operation in hot water supply systems of residential buildings, social and cultural facilities and industrial facilities. Over 1000 m2 of solar collectors have been installed by the company, including Baranovichi AP-2, Belaya Rus state farm (Uzden district), Slutskaya poultry farm (Selishcha village), sanatorium "Belarus" (Sochi), Botanical Garden of NAS of Belarus (Minsk), etc. [5].

The TERMOSOLAR solar collector range (Germany) is the only industrial manufacturer of flat-plate vacuum collectors in the world [6]. THERMO/SOLAR Žiar s.r.o., one of Europe's leading manufacturers of thermal solar collectors, was founded on 1 January 1992. It draws on over 20 years of experience in the development and production of solar collectors of its founders. These are the aluminium mill ZSNP a.s. Žiar nad Hronom and thermo|solar Energietechnik Regensburg - at that time the largest manufacturer of solar collectors in Germany.

The current production capacity is 100,000 m^2 collectors per year. It can be doubled in the shortest possible time. THERMO/SOLAR Žiar s.r.o. is one of the few manufacturers where casing forging, production of selective coating and final assembly of solar collectors are concentrated in one plant. Different types of collector support structures are also produced, and the customer receives all the necessary components for the assembly of the solar plant at one location. In Western Europe, solar collectors from different manufacturers are usually compared by means of the so-called "capacity price", i.e. the cost per unit of heat produced, defined as the ratio of the capital costs to the actual heat gain under certain operating conditions. THERMO/SOLAR collectors always come out on top in such comparisons. Maintaining and even improving this position is the company's primary objective for the coming years.

Benefits of THERMO/SOLAR collectors

- Very good value for money

- the world's only industrial manufacturer of flat-plate vacuum manifolds

- 30 years of experience in solar collector production

- 12-year warranty on collectors, proven in practice to last 25 - 30 years

- large production capacity

The flat-plate vacuum collector, an industrially unique collector of its kind. It is designed for installations that require an operating temperature of more than 80 degrees Celsius or are used in cases where high heat output is required during periods of low solar radiation (winter period). The vacuum guarantees stable parameters throughout the collector's service life.

Designed for installation vertically in forced-circulation solar thermal systems. Consists of a compact pressed casing, to which a special solar safety glass is attached by means of a frame made of stainless aluminium profiles. The absorber sections are made of bent aluminium-magnesium profiles with a highly selective coating, which are attached to a tubular coil made of copper tubes by means of pressing. The manifold leads are connected to the hydraulic and vacuum circuits by mechanical clamps with AElig; 40mm. The manifolds are connected in parallel to each other, in one row max. 8 pieces. The residual gas in the manifolds can be replaced by krypton.

Figure 1.9: TS400 flat-plate vacuum manifold [6].

Flat collector with tubular copper pipe coil designed for installation vertically in forced circulation solar thermal systems. It consists of a compact pressed casing, to which a special solar safety glass is attached by means of a frame made of stainless aluminium profiles. Absorber sections are made of bent aluminium-magnesium profiles with highly selective coating, which are attached to a tubular coil made of pressed copper tubes. The AElig; 18 mm copper tube leads are connected to the hydraulic circuit by means of mechanical clamps. The manifolds can be connected to each other in parallel, up to a maximum of 8 pieces.

Figure 1.10. Operating principle of a solar water heater with a THERMO/SOLAR collector [6].

The heat absorbed by the solar water heater heats the water, and hot water is lighter than cold water, hence it can evaporate and rise to the upper layers through convection. This process takes place between the pipes and the storage tank, resulting in a natural water circulation without any pumping equipment.

The main role of a thermo tank - a heat accumulator tank with a special coating - is to maintain the temperature of the water entering it.

All of the heat absorbed by the solar water heater can be used for the same needs that we are used to solving with a traditional gas boiler. Although a solar heater is slightly more expensive than a boiler, the overall payback period will be no more than a year and a half or two years, primarily due to the savings obtained by not using gas (gas + inflation + boiler maintenance).

The use of a solar heater will enable consumers to significantly minimise their gas costs, as a conventional heating boiler is the main consumer of household gas. More than 80% of its monthly cost is attributable to the boiler (sometimes this figure reaches 95%). This is why installing solar water heating systems can reduce gas consumption by at least 80%!

The tank capacity is calculated according to the number of hot water consumers, e.g. shower/bathroom, domestic use. The average daily consumption of hot water for one person is approximately 50 litres, so for a family of three at least 150 litres is needed.

Fig.1.11. General view of the solar water heater Dachnik [6].

Solar water heater Dachnik - free hot water for 15 seasons.

In addition, the Dachnik can easily be used not only in the summer house, but also in other places with seasonal water consumption - public showers, children's summer camps, etc.

The Dachnik solar water heaters have all the features you need:

- water is heated in a short time by a vacuum collector
- effective both in cooler and cloudier weather
- easy installation and no need for a water connection
- tank capacity available - 90, 150, 200, 250, 300 litres - for any consumption volume

For year-round use, the "Dachnik +" model is used. This model allows you to heat water even at sub-zero temperatures outside and also allows you to use pressurised water (from the water mains). The collector of the Dachnik + heater uses Heat pipe and non-freezing coolant, which ensures its year-round use. The model also comes with a controller, a temperature sensor and a 1.5 kW electric heater.

Figure 1.12. General view of the Universal solar water heater [6].

Warm water is always needed in a private home or cottage. Commercial facilities such as tourism and catering - cafés, restaurants, hotels, boarding houses, holiday homes, tourist and sports complexes - are also in need of it. To wash hands and dishes with warm water, take a shower, a bath or to take a warm pool - all these things have to be paid for. On the scale of any facility, tangible amounts of money are spent on hot water.

The Universal solar water heater can provide up to 90% of these needs, almost entirely covering the cost of hot water. "Universal is designed simply and reliably.

The use of modern technology and materials and the ergonomic design make the operation of the solar water heater pleasant and carefree. And the service life (at least 15 years) will make it possible to forget all the problems associated with hot water for a long time.

The different versions of the Universal can cover the needs of any facility - cottage or small organisation. Possible tank volumes are 150, 200, 240, 300, 360 litres. And the installation of several large units will cover all needs.

The water heater works all year round and at first this is questionable for many people. How can you heat water on a frosty day? In the same way, people didn't believe at first that a car could drive and an aeroplane could fly. But after getting acquainted with its structure and principle of work, and best of all - after seeing it with their own eyes, people abandon their doubts. After all, the efficiency of the technology is such that a water heater can bring water to a boil!

Solar collector Heat pipe

*Figure 1.13. Heat pipe **manifolds**.*

The Heat pipe has the same structure as the Simple pipe - it consists of two tubes inserted into each other, between which there is a vacuum.

As with the Simple tubes, heat loss is minimised, but solar efficiency is close to 100 percent due to the special selective coating of the tube. This means that heating is quick - the temperature starts to rise 2-3 minutes after sunrise.

The Heat pipe, however, has a number of important differences. There is a copper tube inside the vacuum tubes, through which a non-freezing liquid circulates.

This makes the Heat pipe collector efficient at all temperatures, and its operation only depends on the presence of sunlight.

They are unaffected by cold and can work in sub-zero temperatures (down to -50 degrees Celsius)

Heat pipe manifolds are designed for use in closed systems connected to a pressurised water supply.

§1.4 Types of solar water heaters.

Solar water heaters can be of the active or passive type. An active system uses an electric pump to circulate liquid through the manifold; a passive system has no pump and relies only on natural circulation. There are experimental models where the pumping of the heating medium is carried out Stirling pumpThe solar thermal system has a solar thermal system, which receives energy from the sun.

Fig.1.14 Passive water heater.

Passive systems. Passive (Thermosyphon) systems move the finished water or coolant through the system by means of natural gravitation resulting from the difference in density between the heated and cooled thermal fluid. Passive systems with convection are cheaper than active systems, but also less efficient because of the slow circulation in the system. Heat pipe systems are more expensive than convection systems but have lower operating costs. In addition, heat pipe systems allow heat to be pumped downwards, i.e. against the forces of convection. The characteristics are highly dependent on the specific type of pipes.

Active systems. Active systems use electrical **pumps**valves and controllers to circulate **coolant pumps** through the collector. They are usually more expensive than passive systems, but also more efficient.

Active open loop systems

Active open loop systems use pumps to circulate water through collectors. Active open loop systems are popular in regions with positive temperatures or for seasonal use. They can be operated at air temperatures as low as -20 C or -25 C.

Active systems with a closed circuit. In these systems the heat transfer medium for the collector is usually water-glycol antifreeze. The heat exchangers transfer the high temperatures from the coolant The heat exchangers transfer the high temperature of the primary circuit to the water stored in the tanks (thermal storage tanks). Closed loop systems are popular in areas exposed to continuous sub-zero temperatures, as they have good frost protection. Due to the high stagnation temperatures during periods of maximum irradiation, not all antifreezes are suitable for use in solar thermal systems.

Market analysis of the domestic solar collector market. There are only two types of domestic solar collectors widely available on the market - flat-plate and vacuum-tube collectors.

Distribution. World leader in production and application - . **China**. В **2007** China has used solar water heaters for about 40 million **families** with a total population of 150 million. By 2009, the total area of installed solar water heaters had increased to 140 million m². This is enough to supply hot water to approximately 60 million households. К **2020** 300 million m² of premises in **China** will be equipped with solar water heaters.

Figure 1.15. Solar water heaters installed on the roofs of many new houses in a Chinese province <u>Hubei</u> [7].

Water heaters are also very widely used in IsraelThe country has a water heater, where 95% **of** the flats are equipped with such appliances. This is due to a law passed in 1976 and obliges dwellings to be built with integrated solar water heaters. The exception is some high-rise buildings (over 24 storeys), where the roof area is insufficient to accommodate sufficient solar collectors for all consumers in the building. This widespread use of solar water heaters saves around 4% of all electricityproduced in the country.

Figure 1.16. Modern Israeli homes equipped with solar collectors [7].

In the west of the city of Kreilsheim (Germany), [7] instead of traditional fuels, locals are increasingly using renewable energy sources to meet their heating and hot water needs (http://www.rehau.ru). It is here that a modern hot water generation system has been applied, using a 10,000 square metre solar collector. The southern part of the 7,500 square metre collector, located on a special 15 metre high earth berm, has taken on the lion's share of heat production from solar energy for the 2,000 inhabitants. The solar plant is supplemented by seasonal heat storage using geoprobes.

First, the heated water from the collectors flows into a 480 cubic metre buffer storage tank. Since most of the solar heat is produced in the summer months and is mainly used for heating during the cold winter period, the excess heat must be stored somewhere. One of the largest geofence heat storage facilities ever built in Germany is being used for this purpose. It consists of 80 RAUGEO geoprobes 55 metres deep and makes it possible to store the summer heat in the ground until winter. The probes are made of cross-linked polyethylene (PE-Xa) and can withstand temperatures of up to 75 °C.

The solar thermal system in Hirtenwiesen (Germany), [7] produces 3 million kilowatt hours of thermal energy annually. The supply of heat to the companies in Kreilsheim from a solar plant reduces the need for conventional fuels by half and frees the environment from 1,000 tonnes of carbon dioxide emissions per year. The "District Energy Initiative", which is supported by the federal government, awarded the installation the "Lighthouse Project" title. It is awarded to innovative projects with future-oriented, environmentally-friendly technologies.

Solar shower installations. Solar collectors are recommended for use in domestic hot water installations: in garden cottages and shower installations, in detached manor houses. The use of one solar collector in the shower cubicle of a garden cottage saves more than 1 m^3 of firewood per season, and the use of solar collectors for domestic hot water supply in a one-family three-room residential building saves up to 5 m^3 of firewood per season or more than 1 ton of coal [8].

Otler (UK) [9] is a modern, premium shower unit. Tropical showers are huge shower heads with a large number of nipples; the water flowing out of them gives a massage effect and a lot of pleasant sensations. Colour LEDs embedded in the shower surface illuminate the water jets; this effect is called 'chromotherapy' - the positive effect of colour and light on the human body - and has long been studied by psychologists.

But technology has not stood still and Otler now produces five new models of rain showers that do not require electricity - Altair, Altair Grand, Antares, Isida and Lira.

In the new Otler range, the mechanical energy of the water flowing through the shower panel is converted into electrical energy to power the LEDs.

This greatly simplifies the installation and operation of the showers, as there is no need for a separate power supply to operate the product.

The chromotherapy principle has also been improved, as the light from the diodes now varies according to the temperature of the water. Hot water has a red light, while cold water has a blue light. Now that you can tell the temperature of the water by the colour of the diodes, you won't get burned when you turn on the shower.

In addition, by eliminating unnecessary electrical supply elements, the cost of the shower panels has been reduced significantly, while maintaining the same quality and increasing comfort of use.

Shower enclosures from leading manufacturers [9]. Shower enclosures are becoming increasingly popular because of their convenience and compact size. No room is without a shower enclosure - whether it's a home, an office or a restaurant. We got so used to it that we stop noticing it, but when it becomes necessary to buy a shower cabin or acrylic bath, it can turn into a real problem. After all, there are many shops that sell sanitary ware - both specialised and generalist. It can be difficult to find your way through all the variety.

Istok specialises in the sale of shower enclosures in St Petersburg.

-Shower cubicle 7080 is a minimum of comfort at a low cost,

-standard showers are the best combination of price and quality,

-premium showers with a wide range of functions.

-shower cubicles with sauna, shower enclosures.

Fig.1.17 Shower enclosures from leading manufacturers Appollo A-0820 shower enclosure [10].

Traditional, simpler and cheaper shower enclosures with solar collectors. Beach shower enclosures with solar collector heating. They do not require electricity and/or gas to heat water and have an integrated 130 litre hot water storage tank. They have an attractive design. Environmentally friendly and self-contained source of hot water.

There are two possible layouts for such systems:

1. With tank with heat exchanger.

2. With tank without heat exchanger.

The main technical characteristics of the installations:

RKFD-1558-130 manifold

Number of tubes, mm - 15

Tube diameter, mm - 58

Storage tank capacity, l - 130

Peak output, kW - 2.32

Daily production of hot water, l - 500

RKFA-1847-100 manifold

Number of tubes, mm - 18

Tube diameter, mm - 47

Storage tank capacity, l - 100

Peak output, kW - 2.13

Daily production of hot water, l 450

With all the possibilities of modern household appliances in the countryside, the pleasure of taking a summer shower cannot be denied. The simple outdoor shower cubicle design with a solar thermal tank on top can be upgraded to the point where hot water is available around the clock.

In a normal summer shower, the sun's rays heat the water in the tank. In southern areas, on a clear sunny day, it can get as hot as 50-60 °C. This is more than enough for a hot shower, but overnight the water in the tank can cool down to ambient temperature, especially if the nights are cold. In this case, even a warm shower in the morning is out of the question.

Water in the tank could be kept hot by insulating it, but then the tank would also be insulated from the sun. The shape of the tank and the material from which it is made play a certain role in the efficiency of heating. Metal tanks are preferable to plastic ones, because metal conducts heat much better. If you make the tank airtight, the water will expand when heated and create additional pressure inside it, which can be used as a factor to increase the head of water, when taking a shower. In this case, however, the walls of the tank must be strong enough to withstand this pressure. Water by nature cannot compress under pressure, and if the volume exceeds the capacity of the tank, the tank is likely to fail and crack. To prevent this from happening, 15% of the tank volume must be left unfilled, then air will act as a shock absorber as it is able to compress to a certain degree.

However, do not get hung up on the idea of extra pressure in the tank. In practice it discharges quite quickly and when the water level in the tank reaches half of its level, the excess will disappear. If the level drops, it will be necessary to depressurise the tank so that the water can flow out by gravity in peace. For this purpose, air check valves are used, which open when the pressure inside the tank is negative.

So, in order to make an ordinary shower created according to the above described scheme - a heat-saving shower, it is necessary to make some additions to its construction. One of the methods implies the use of an additional insulated tank and a water pump. The operation of such a system is as follows:

- cold water flows into the water heating tank;

- when the tank is full, the float valve shuts off the cold water;

- A temperature sensor installed inside the solar tank activates when the control temperature (35-60°C) is reached and closes the circuit, opening the solenoid valve to drain the water into the insulated tank;

- Once the solar tank is empty, the solenoid valve closes and the cold water valve opens;

- The "solar tank" is filled with another batch of water and then everything happens in the same order.

When the shower is running, hot water is supplied by a water pump or by gravity if an insulated tank is installed above the showerhead level. The electric water pump consumes a small amount of electricity, which is nothing compared to the cost of heating water with commercially available energy. The advantage of the described system is that the water accumulation in the thermally insulated tank is automatic. However, even in a "gravity" system (in which the heating tank is located above the thermal insulating tank), a water pump is essential. It will pump (at the signal of the temperature sensor) the unused water from the thermally insulated tank into the heating tank, resulting in fast heating of the cooled water as well as saving cold water.

A thermally insulated tank is a rectangular or cylindrical plastic tank insulated with one of the thermal insulation materials (foam, mineral wool, etc.).

§1.5 Key findings and conclusions based on an analysis of the research and development of modern solar water heaters

The following conclusions can be drawn from an analysis of the design, performance and economics of modern solar water heaters:

1. All solar collectors are conventionally divided into flat-plate collectors and vacuum collectors (on vacuum tubes).

2. Flat panel solar collectors are an absorber, an element that absorbs solar radiation and is connected to a heat conducting system. On the outside, the element is covered by a layer of transparent material, a transparent coating. This coating is most often made of special tempered glass, in which the metal content is reduced as much as possible. The reverse side is closed with a thermal insulator to reduce heat loss. If the heat is not transferred to external consumers, such a flat-plate collector is able to heat the intermediate heat carrier up to one hundred and forty degrees. Special optical sheaths are currently being developed and used. Since copper has the highest thermal conductivity of all the materials used, it has become the main raw material for producing the absorber.

Flat-plate collectors. The copper tubes of the absorber of a flat-plate collect the solar heat under the protective glass, which creates a greenhouse effect, and transfer the heat to the heat transfer medium or water circulating in the tubes. The mathematical modelling of the elementary solar water-heating installation spent at Institute of high temperatures of the Russian Academy of Sciences with use of modern software and the data of typical meteoyear has shown that in real average climatic conditions use of seasonal flat solar water heaters working during the period from March till October is expedient. For an installation with a solar collector area to storage tank volume ratio of 2 m^2 /100 l, the probability of daily water heating during this period to a temperature of at least 37 C is 50-90%, to a temperature of at least 45 C - 30-70%, to a temperature of at least 55 C - 20-60%. The maximum probability values refer to the summer months.

Advantages and disadvantages of flat-plate collectors. The main plus point is that, in terms of maintenance, it is completely trouble-free. Flat-plate collectors are well suited even for humid climates. They work better in cloudy weather.

Cons. Firstly, their efficiency is lower than that of vacuum collectors. Secondly, flat-plate collectors (European) will cost more than vacuum collectors (Chinese).

3. In the vacuum cylinder water heater, the volume containing the surface that absorbs solar radiation is separated from the environment by an evacuated space, which virtually eliminates heat loss to the environment by means of conduction and convection. Radiation losses are largely suppressed by a special selective coating. As the total loss factor in the vacuum collector is very low and the heat transfer medium can be heated in the vacuum collector to temperatures of 120 160 °C. The solar vacuum collector enables the collection of solar radiation in all weathers, almost regardless of the outside temperature. The energy absorption coefficient of these collectors can be up to 98 %. Advantages and applications of vacuum solar collectors. Due to their high thermal insulation, vacuum solar collectors are very efficient at low ambient temperatures. Solar thermal systems based on vacuum collectors can be used both for hot water supply and for home heating. In this case in the summer time it is possible to receive hot water completely from the solar heater. During the rest of the year, up to 80 % of the hot water can be provided by solar energy.

The main part in vacuum manifolds is a special vacuum tube, coated with a niello for heating, which contains water or antifreeze. The whole construction is made according to the principle of a thermos device. Around the cavity filled with fluid a kind of vacuum chamber is created to reduce non-productive heat losses. Using such element it is possible to heat water even if the ambient temperature is minus. And to meet the demand for hot water by 60 percent. In order to increase the efficiency of devices, the internal vacuum tubes are made in a faceted shape or in the form of the letter "U". The outer shell of the tubes is made of borosilicate glass, which has increased strength and does not lose its optical properties for a long time. The use of systems built on vacuum solar collectors can provide the population with a third of the energy required for heating in autumn or spring.

Solar collectors equipped with heat pipes (vacuum tube Heat Pipe or "heat pipe") have been spreading lately. Inside such a tube is a liquid that has a reduced boiling point, e.g. ammonia. One end of the tube is inserted into a heat exchange tank. When heated by solar radiation, the liquid boils and the steam rises up and transfers the heat to the heat transfer medium circulating in the common collector. Solar collectors with such tubes are more efficient than any other collector. In addition to increased efficiency, it is also extremely resistant to mechanical influences.

Advantages and disadvantages of vacuum collectors. Pros - this equipment is cheaper and has a higher efficiency, especially when the weather is clear. Cons. Practice has shown that the vacuum collector does not handle humid climates well. In cold seasons, when frost falls on its surface, the collector can not capture the sun's rays and stops working, you will need to monitor this and, if necessary, clean off the frost (like a car windshield) to make the system work again. But this requires an optimum ratio of humidity to air temperature, which of course happens, but not so often as to be fatal.

4. Solar collectors of different types provide thermal energy that is primarily used for hot water preparation, which is especially relevant during the summer season when there is maximum solar activity and maximum hot water consumption.

There are two main types of collectors - flat-plate and vacuum collectors. Vacuum collectors are the more complex ones. On sunny summer days, the difference in the operation of good flat-plate and vacuum solar collectors is hardly noticeable. During the warmer months, excess heat is generated that needs to be used, e.g. for swimming pools, showers, laundry, etc.

5. Normally, solar collectors are installed fixed, and the angle of inclination is chosen according to the main purpose of the unit. When installing, the collector should be oriented towards the south, but always with reference to the topography of the terrain. It is advisable to deviate from the southern orientation by no more than thirty degrees, so that the heat will also be generated within the limits. In northern areas it may be of interest to install at an angle close to vertical, when the receiver will make more use of the rays of the low-lying sun, including those reflected from the surface of the snow, and heat production during this period will be maximum.

6. Two flat-plate collectors (e.g. VITOSOL 200-F) with a total surface area of 4.6 m^2 or one vacuum collector with a total surface area of 3 m^2 (e.g. VITOSOL 200-T) are usually sufficient for a family of three to four people. The collector surface area is dimensioned in such a way that 50 to 60 % of the total heat demand for heating hot water is covered by heat from the sun during the year. The excess heat can be discharged into the heating system. The use of solar heat for heating purposes would require a much larger collector area. Using solar collectors can reduce hot water heating costs by 66 % and heating costs by 30 % per year. The use of solar thermal systems reduces heating and hot water costs considerably. Reduction of operating costs. Longer service life of the auxiliary heating system.

Under Uzbek conditions, the energy income from 1 m^2 of collector is at least 1000 kWh/m^2 per year. The system can provide hot water for 300 days a year.

In summer, the consumer will be able to make do with solar collectors only, while in winter, the two systems, i.e. a doubler (gas or other) and a solar collector, will functionally complement each other.

The efficiency of solar domestic hot water heaters lies in saving fossil fuel or electrical energy through the use of solar energy. Example. Let's assume that the average gas-fired water heating in a small hotel for 60 people costs 1,600,000 UZS per year. Solar water heating systems can save 50-65% of the cost per year. Modern solar domestic water heaters can save annual energy costs of up to 60% of the energy needed to heat water. In Uzbekistan, properly installed systems meet 95% of the heat demand between March and November.

7. The production of solar thermal systems has increased several times over the last 4 years worldwide. Today more than 3 million plants are commissioned per year. Vacuum solar collectors have proven their efficiency even in the climatic conditions of Alaska. This rate of uptake of this type of energy has been made possible by such factors in recent years as:

- development of solar collectors with vacuum-type tubes that allow operation at sub-zero temperatures of -30°C and even -50°C, which is relevant for winter conditions;

- cheaper collector production thanks to the transfer of production from the world's leading manufacturers to the Asia-Pacific region;

- higher energy prices.

8. At present, a psychological factor prevents the widespread use of solar collectors in Uzbekistan. Subconsciously the solar energy is not considered constantly accessible because of weather and seasonal conditions though technically this problem is solved very simply by use of heat accumulators. Inertia of thinking, including specialists. Old style heating systems are installed out of habit. Initial costs are higher than classical heating systems. In the near future there will be a rapid inflation of this market. Further increases in energy prices in Uzbekistan will force consumers to use new efficient solar heating and hot water systems.

9. The service life of most solar plants designed and installed by highly qualified professionals can produce heat for more than 30 years.

10. The operating and maintenance costs of modern solar domestic water heaters are relatively low. The units operate automatically. A periodic inspection by a specialist is all that is required. The antifreeze coolant must be replaced every

six years or so. The manifold withstands rain, hail and frost. In winter, an additional de-icing system can be installed. The glycol contained in the collector withstands temperatures down to -50 °C.

Of the modern collectors, German collectors do not require servicing, whereas Chinese collectors require servicing every 2-3 years. However, Asian products are comparable to boiler equipment in terms of cost, and European products are much more expensive. At the same time, German collectors have a service life of 25 years, and they pay for themselves in 10-15 years.

11. One m² solar collector for a domestic water heater costs \$320-\$330. But one collector without basic and additional water equipment will not heat anything. The calculation of cost of modern foreign systems for heating of water for a family from 4 persons (water consumption in day - 100 litres) - 2 collectors, a boiler, an expansion tank, the control system, accessories will cost \$4 000-4 500. It should be borne in mind that each system is individual, and the percentage of energy savings when using a solar thermal system must be calculated. For the calculation of solar thermal systems, special sophisticated programmes are used.

In summary, the following conclusions can be drawn from a review of the design, performance and economic features of modern solar water heaters:

- The advances in the development of solar water heaters are quite high, so they are already being used by individual consumers.

- However, modern water heaters are structurally complex and quite expensive, i.e. economically inefficient for the average consumer,

- The widespread use of solar water heaters in Uzbekistan requires more engineering and economic research, the development of design and practice, and the commercialisation of these at an affordable price for individual consumers.

§1.6 Design requirements for a domestic solar water heater

1. The developed design of the conical-type solar domestic water heater must meet the requirements for conventional domestic appliances, i.e. it must be easy to carry, small in size and weight, aesthetically pleasing and harmonious in relation to surrounding objects, and durable and cheap.

2. The design of the solar water heater should be developed on the basis of worldwide experience in improving and cheapening the design development of the currently most demanded solar domestic water heaters.

3. The solar water heater must consist of the following components: a storage tank-collector, a shower enclosure or a frame for installing the storage tank, and water mixers. All these parts and materials must be domestically manufactured and comply with the requirements of state and standards.

4. The storage tank-collector of a domestic solar water heater must be made in the form of a truncated cone of galvanised sheet iron, allowing a considerable simplification of the design and manufacturing technology of the water heater, reducing metal intensity and ensuring lightness, compactness and ease of operation.

5. The angle between the forming surface of the absorbing surface and the base of the truncated cone must be no greater than 35°, which ensures maximum illumination of the entire side surface of the cone painted black and the greatest absorption of solar energy. Moreover, an angle of 0.9φ (φ is the latitude angle of the terrain) is considered optimum for spring and summer operation.

6. In order to significantly increase the efficiency of the collector area and practically make the stationary collector surface of the water heater traceable, the absorption surface must be made in the form of a truncated cone, as the surface of the cone is constantly directed towards the sun.

7. The heater must be protected against the damaging effects of the external environment (enamel coating, impregnation, sealing, etc.).

§1.7. Performance requirements.

The main parameters and dimensions of the solar water heater should be as follows:

1) Solar water heater storage tank capacity, l, at least-110.

2) absorption surface area, m^2 -1.5

3) The daily capacity of the water heater shall be at least 110 litres.

(the water heater capacity is taken on a clear sunny day at a water temperature of 37 C during the spring-summer season from May to August for central regions and from April to September for southern regions
 with the hotter water diluted to this temperature)

4) The maximum water heating temperature (depending on the level of solar radiation) is $37-60^0$ C.

5) The domestic cone-type solar water heater is available in two installation options:

type 1 - on the frame;

type 2 - on a cabin frame (shower room)

Both must have a mixer tap with a spout and a shower screen.

6) The dimensions of the water heater shall be 1000x1000x2000 mm.

7) The operating pressure of the heater must be equal to the hydrostatic pressure of the water filling it.

8) Total heat loss coefficient, W/m K, not more than 9.

9) optical efficiency, %, at least-75.

Regardless of the configuration of the product, the heater itself is tested without the mixer.

Test requirements.
Acceptance tests of the water heater are carried out for:

1) determining whether the prototype to working design documentation meets the requirements of the specification;

2) The decision as to whether or not a product should be put into production.

Personnel who have studied the water heater specifications and have been instructed in safety must be allowed to carry out the test.

The tests must be carried out using:

1) tape measure ZDK2-5ANTD0 GOST 7502-80;

2) 3/4", 1/2" threaded rings GOST 18U2U-73;

3) scales with a measuring range of at least 100 kg and an accuracy of 0.1 kg;

4) pressure test bench;

5) thermal-hydraulic test bench;
6) performance test bench.

An overall assessment of the quality performance of the products as a result of the tests.

Based on the results of the tests, an overall assessment is given of the quality performance of the products and compliance with the requirements of the technical specification in the quality category (reliability and serviceability of the design, performance, etc.)

Performance requirements.

1. The heater must be designed to be easy to operate, easy to inspect and check, serviceable by personnel of normal qualifications, lightweight and transportable.

Maintenance and warranty requirements.

1. The design of the solar water heater should not require maintenance during operation other than periodic cleaning of the storage tank-collector and replacement of the mixers.

2. The warranty period for the solar water heater shall be 3 years from the date of sale through the retail network. Within the warranty period, the manufacturer undertakes to repair all faults free of charge, provided that the owner complies with the rules of operation, transportation and storage. The right to warranty repair is secured by a warranty card.

CHAPTER II

DESIGN, MANUFACTURE AND TESTING OF A SOLAR CONICAL SHOWER WATER HEATER VB-CT

Conical domestic solar water heater VSB-KT. OT-ID/11-4-55 (hereinafter referred to as water heater) is designed for obtaining hot water and showering in spring-summer-autumn period by solar radiation without using any traditional energy source in the central and southern regions of the country.

§2.1. Design and principle of operation of a solar water heater-shower cone.

- A solar water heater works by converting solar radiation energy into thermal energy, transferring this energy to water and storing the heated water for later use.

- General view of the conical type shower water heater VSB-KT. OT-ID/11-4-55 is shown in Figure 2.1. The solar water heater consists of a hot water storage tank-collector, a transparent membrane cover, a mixer, a frame (type 1) or a cabin-shower frame (type 2) for mounting the storage tank-collector, and a set of piping, connectors and fasteners (Fig. 2.1).

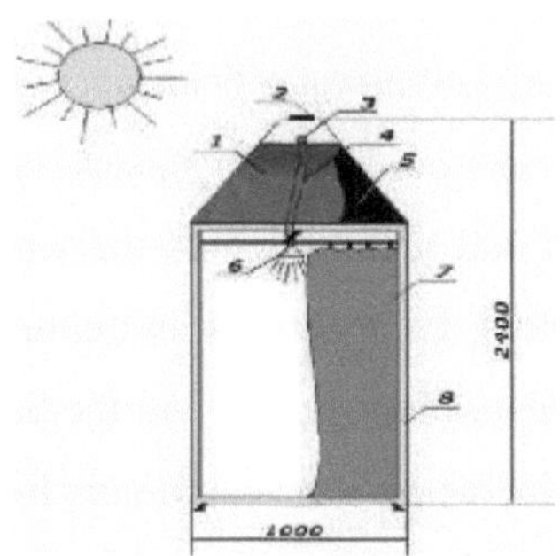

Fig.2.1. General view of solar domestic water heater of conical type: 1 - collector of solar energy in the form of a conical shape accumulator tank; 2 - accumulator tank lid; 3 - deflector; 4 - rubber hose; 5 - float; 6 - transparent cover-insulator; 7 - valve; 8 - shower.

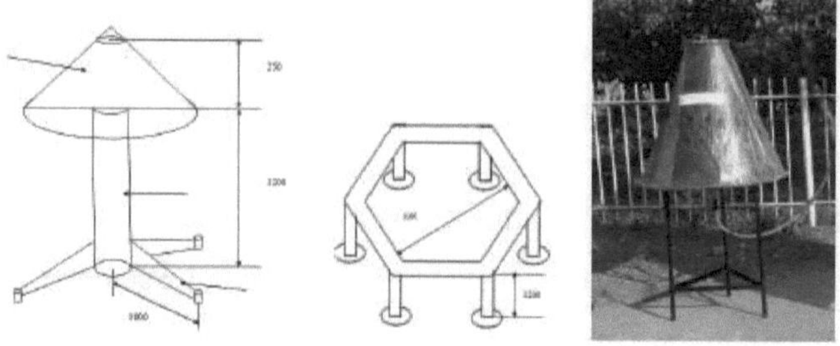

Fig. 2.2. Frame options for installing a solar domestic hot water heater - tapered shower enclosure.

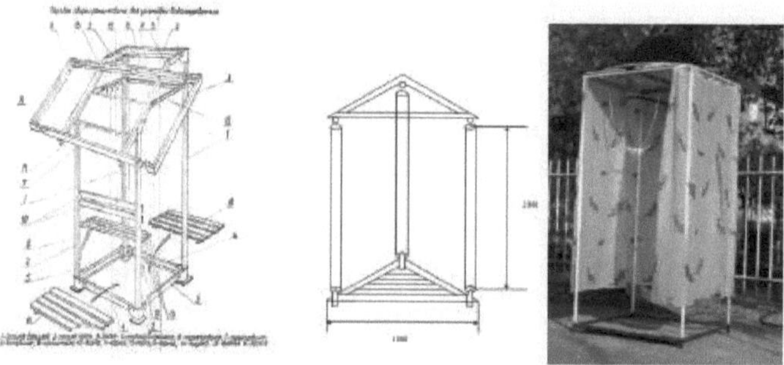

Fig.2.3 Frame options for installing a solar domestic water heater - tapered shower enclosure.

The storage tank-collector (Fig. 2.4) of the domestic solar water heater is made in the form of a truncated cone of galvanised sheet iron. The galvanised sheet iron cylinder-collector tank simplifies the design and technology of the water heater, reduces metal intensity and ensures lightness, compactness and ease of operation. The angle between the forming of the absorbing surface and the base of the truncated cone is not more than 35 °, which provides maximum light painted in black entire side surface of the cone, as well as the greatest absorption of solar energy. Moreover, an angle of 0.9φ (φ being the latitude angle of the terrain) is considered optimum for spring and summer operation. The execution of the absorption surface in the form of a truncated cone significantly increases the efficiency of the collector area and practically makes the stationary collector surface of the water heater traceable, as the cone surface is constantly directed towards the sun.

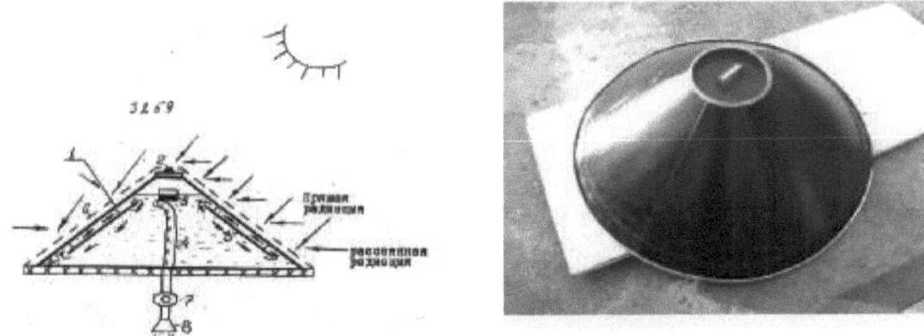

Fig.2.4 Tank-storage-collector.

Fig. 2.5 Water heater frame - (type 1).

The heater frame (Fig. 2.5) - (type 1) is designed for mounting the storage tank and the piping system mixer on it. This simplifies the design as much as possible and reduces the cost of the heater. To make it cheaper (Fig. 2.4), it is possible to exclude the deflector 3, transparent cap-insulator 6, and even the hot water extraction system, i.e. the shower head 8, valve 7, float 5, and rubber hose 4. The hot water is taken from the top cover of the tank manually. If hot water remains in the tank at the end of the day, cover the tank overnight with any insulating cover (felt, old blanket, etc.), then the tank will have a hot water reserve in the morning.

The accumulator-collector tank (figure 2.4) is fitted on the inside with a deflector to promote circulation and a float with a rubber hose to ensure that heated water is drawn from the top of the tank. Hot water from the tank is withdrawn through a valve and shower.

Figure 2.5. Photograph of an experimental prototype of a conical-type solar domestic water heater.

The shower enclosure frame (Fig. 2.6) - (type 2) is a single shower enclosure with a collector tank at the top and a screen curtain. For easy transport, the shower enclosure is available in two demountable versions.

a) assembled b) disassembled

Figure 2.6. Shower cubicle frame. Fig.2.7. Triangular shower enclosure frame.

a) assembled b) disassembled

Fig.2.8. The cab frame of the quadrangular type.

The heater is designed with a mixer from a Sm-K water heater, which permanently connects the internal volume of the heater to the atmosphere and thus relieves the heater from the pressure of the water mains.

When the heater is mounted on a frame or cabin frame, the tank-storage-collector is piped together to form a single closed hydraulic circuit that is filled with water.

To ensure optimum water heater conditions, the storage tank-collector must be oriented strictly towards the south to get maximum light exposure during the day.

A test of the water heater under natural conditions confirms the technical characteristics and the possibility of using them for hot water supply to individual consumers.

The water heater works as follows (figure 4.2.1). Storage tank 1 is filled with cold water. Solar radiation passes through the film transparent hood 6 and is absorbed by the surface of the conical tank 1, painted with black paint. Solar radiation energy is converted into thermal energy, which is transferred to the water that fills the tank. The heated water gradually collects in the upper part of the tank as a result of natural circulation. In this way, the entire volume of the tank is heated during the day.

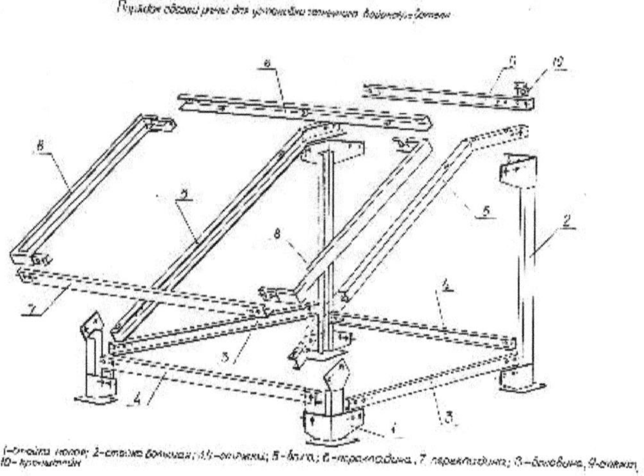

Fig.2.9 Frame assembly procedure.

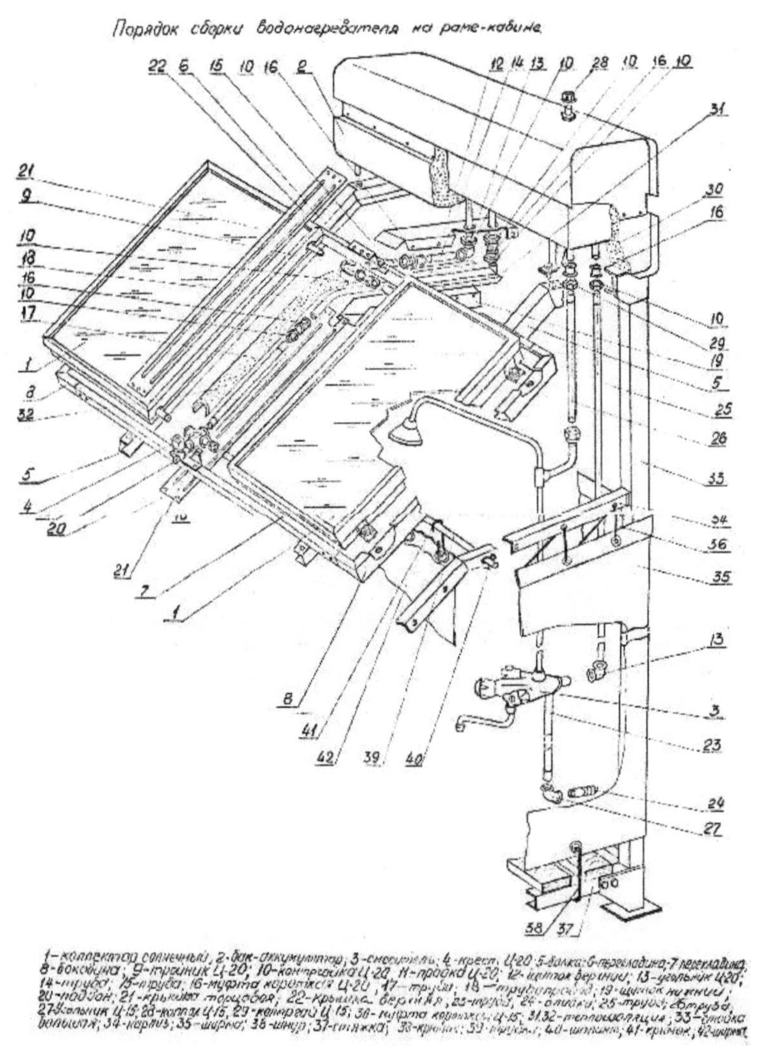

Figure 2.10 Alternatively, the solar domestic water heater can be made into a shower heater.

§2.2. Technical and operational performance of the VSB-KT water heater.

1. solar water heater storage tank capacity, l, no less-110.

2. The daily capacity of the water heater should be, l, at least-110 (160 -250) (the bracket indicates the capacity of the water heater taken on a clear sunny day at a water temperature of 37 C during the spring-summer season from May to August for central regions and from April to September for southern regions, taking into account diluting hotter water to this temperature).

3. maximum water heating temperature (depending on solar radiation) is $37-60^0$ C.

4. The domestic hot water heater of the conical type is available in two installation variants:

type 1 - on the frame;

type 2 - on a cabin frame (shower room).

Both must have a mixer tap with a spout and a shower screen.

5. Water heater dimensions-1000x1000x2000 mm.

6. The operating pressure of the water heater is equal to the hydrostatic pressure of the water that fills it.

7. Weight of water heater without water, kg, max:

type 1 -25.

type 2 -50.

8. Total heat loss coefficient, W/m K, max. 9.

4.3.9 Optical efficiency, %, not less than-75.

§2.3. Test programme and methodology for the VSB-CT conical solar domestic hot water heater

- Determination of the performance of a cone-type solar domestic water heater under Ferghana Valley conditions.

- Demonstration of the conical type solar domestic water heater to possible customers and collecting their feedback.

Test requirements

The following requirements apply to the tests:

- Prepare a site for the installation and installation of the water heater. The water heater is installed on an open, level and unshaded area all day long, provided with plumbing and equipped with a drain for the disposal of used water,

- connect the product to the water supply.

Measurable values

The test process directly measures

- cold water temperature $t_{х.в}$., C

- average temperature of the i-th portion of heated water - t_i C

- the volume of the i-th portion of heated water - V_i

- flux density of total solar radiation falling on the collector surface - E, m/s

- wind speed - V, m/sec

The test process also determines the qualitative characteristics of the sky according to the weather station:

1. cloudless, clear;

2. it is cloudy, sunny;

3. variable cloudiness;

4. Cloudy with clearing;

5. cloudy.

Accuracy of measured values

Temperatures shall be measured with a thermometer with an upper measuring range of + 100 C and an accuracy of 1 C of +-0.5 %,

2. The volume of water heated shall be measured in a calibrated container with a capacity of at least 10 litres and at least 15 litres with an accuracy of + - 1 %,

3. The flux density of total solar radiation shall be measured with a pyranometer and galvanometer type GSA-1 with an accuracy of 1.5%,

4. wind velocity shall be measured with a bowl type MS-13 anemometer with an accuracy of 6%

Test conditions and regimes

1. Preparation of the solar conical type domestic hot water heater for the test consists of filling it with cold water from the water mains. The appliance is

considered filled with water when water flows from the hot water supply pipe to the user. After that, close the valve installed on the said piping and the installation is ready for start-up.

2. Start-up of the solar conical type domestic hot water heater is carried out by removing the protective screens from the glazing of the solar collectors. The start-up takes place between 8.00 and 9.00 local time.

3. The daily output of the conical-type solar water heater is determined between 18.00 and 19.00 local time by draining the heated water from the storage tank. The water is discharged in 10 litre portions into a calibrated container. Use a thermometer to determine the average temperature of each batch of water to be drained. Drainage is stopped when the average temperature of the next batch is less than +37 C

4. The temperature of the cold water that has filled the storage tank while draining the water heated during the day is then determined. To do this, 10 litres of water from the pipeline are poured into a calibrated container and its average temperature is determined using a thermometer.

5. Based on the results of each day's test, a measurement report is drawn up for the daily performance of the conical-type solar domestic hot water heater in solar collector operation.

§2.4 Scope of tests and calculation formulas for processing experimental data

1. At the time of testing, the total operating time of the BSV-CT in solar collector operation must be at least 2.5-3 months.

52. Determining the daily capacity of the BSV-CT when operating with solar collectors consists of determining the actual volume of water heated $V_{факт}$. Conditional, reduced to plus 37°C volume of heated water $V_{усл}$.

3. The actual volume of water heated is determined by the formula

$$Vфак = n \cdot Vi,$$
where

n is the number of portions of heated water with a temperature of at least plus 37°C;

Vi is the volume of the i-portion of heated water discharged during determination of the daily capacity, l.

5.6.4 The nominal volume of water heated to plus 37°C is determined according to the formula

$$Vусл. = \sum Vусл.i,$$

$\sum Vусл.i$ -conditional, reduced to plus 37°C, volume of heated water in the i-portion drained when determining the daily capacity of the BSV-CT, l;

5.6.5. The conditional, adjusted to plus 37°C, volume of heated water in the i-portion drained when determining the daily capacity of the BHW-CT is determined by the formula:

$$Vусл.i = Vi \, (ti - t \, х.в.) \, / \, 37°C$$
where

Vi is the volume of the i-th portion of heated water discharged when determining the daily capacity of the WWTP, 1;

ti - average temperature of the i-th portion of heated water discharged during determination of the daily WWTP capacity, °C; t x.c - temperature of cold water from the water supply system recorded during determination of the daily WWTP capacity, °C.

§2.5. Results of BSV-CT tests at the Ferghana heliopolygon in Ferghana

In April/November 2011-2012, the BSV-CT was set up and tested in-situ at the heliopolygon of the Fergana Polytechnic Institute.

The results of the tests are documented in a report measuring the daytime performance of the BSV-CT in solar collector operation, the results of which are shown in the table.

Measuring the daily output of a solar domestic water heater - the shower cone type.
Place of testing - Fergana
Date 29 August 2011
Qualitative characteristics of the sky cloudless, clear
The cold water temperature in the water mains is 18^0 C.

Table 1

n.p.p. portions of hot water to be drained	The average temperature of the i-th portion of heated water is $t_{i,}$ ^{0} C	volume of the i-th portion of heated water $V_{i,}$ 1	Conditionally adjusted to 37^0 C volume of hot water, $V_{усл.}$	Solar radiation and environmental parameters			
				Time, h	Summarise the radiation drop on the perpendicular. In. W/m^2 , E	wind speed V, m/sec	ambient water temperature t_0 ,^{0} C
1	54	10	18,94				
2	53	10	18,3	9^{30}	745	2,5	18
3	52	10	17,9				
4	51	10	17,37	12^{30}	915	3,0	24
5	49	10	16,31				
6	48	10	15,78				
7	46	10	14,8				
8	44,5	10	14	15^{30}	725	3,1	26
9	42,5	10	12,91				
10	40,5	10	11,75				
11	38	10	10,53	18^{30}	410	3,0	23

$Y_{фак}$ = 110 litres, $Y_{усл.}$ = 180.0 litres.

Measuring the daily output of a solar domestic water heater - the shower cone type.

Place of testing - Fergana

Date 5 September 2011

Qualitative characteristics of the sky cloudless, clear

The cold water temperature in the water mains is 18^0 C.

Table 2

n.p.p. portions of hot water to be drained	The average temperature of the i-th portion of heated water is $t_{i,}{}^0$ C	volume of the i-th portion of heated water $V_{i,}$ 1	Conditionally adjusted to 37^0 C volume of hot water, $V_{усл.}$	Solar radiation and environmental parameters			
				Time, h	Summarise the radiation drop on the perpendicular. Po w/m. W/m^2 , E	wind speed V, m/sec	ambient water temperature t_0 ,^{0} C
1	53	10	18,94				
2	52	10	18,3	9^{30}	760	2,0	19
3	49	10	17,9				
4	48	10	17,37	12^{30}	830	1,0	24
5	46	10	16,31				
6	45	10	15,78				
7	43	10	14,8				
8	42	10	14	15^{30}	690	1,5	28
9	41	10	12,91				
10	40	10	11,75				
11	38	10	10,53	18^{30}	400	1,5	24
$Y_{фак}$ = 110 litres, $Y_{усл.}$ = 160.0 litres.							

Measuring the daily output of a solar domestic water heater - the shower cone type.

Place of testing - Fergana

Date 17 September 2011

Qualitative characteristics of the sky low clouds, clear

The cold water temperature in the water mains is 18^0 C.

Table 3

n.p.p. portions of hot water to be drained	The average temperature of the i-th portion of heated water is $t_{i,}{}^0$ C.	volume of the i-th portion of heated water $V_{i,}$ 1	Conditionally adjusted to 37^0 C. Volume of hot water, $V_{усл.}$	Solar radiation and environmental parameters			
				Time, h	Summarise the radiation drop on the perpendicular. In. W/m^2 , E	wind speed V, m/sec	ambient water temperature t_0 ,^{0} C.
1	55	10	19,47				

2	54	10	19	10^{30}	780	2,0	20
3	52	10	18,68				
4	50	10	18,0	12^{30}	900	2,3	24
5	48,5	10	17,0				
6	47	10	16				
7	46	10	15,5				
8	44	10	15,1	15^{30}	720	2,0	27
9	42,5	10	13,8				
10	39	10	11,8				
11	37	10	11,0	17^{00}	410	3,5	23

$Y_{\text{фак}}$ = 110 litres, $Y_{\text{усл.}}$ = 178 litres.

Measuring the daily output of a solar domestic water heater - the shower cone type.

Place of testing - Fergana

Date 7 April 2012.

Qualitative characteristics of the sky low clouds, clear

The cold water temperature in the water mains is 18^0 C.

Table 4

n.p.p. portions of hot water to be drained	The average temperature of the i-th portion of heated water is $t_{i,}$ ^{0} C	volume of the i-th portion of heated water $V_{i,}$ 1	Conditionally adjusted to 37^0 C volume of hot water, $V_{\text{усл.}}$	Solar radiation and environmental parameters			
				Time, h	Summarise the radiation drop on the perpendicular. In. W/m^2 , E	wind speed V, m/sec	ambient water temperature t_0 , ^{0} C
1	51	10	17,36	9^{30}	850	1,5	9,3
2	51	10	17,36				
3	50,5	10	17,1				
4	47	10	15,26	12^{30}	920	1,2	21
5	47	10	15,26				
6	46	10	14,73				
7	45	10	14,2	15^{30}	890	0,5	21,6
8	43	10	13,16				
9	41	10	11,57				
10	39	10	11,05				
11	37	10	10	18^{30}	670	1,5	19,3

$Y_{\text{фак}}$ = 110 litres, $Y_{\text{усл.}}$ = 157 litres.

Measuring the daily output of a solar domestic water heater - the shower cone type.

Place of testing - Fergana

Date 14 April 2012

A qualitative characteristic of the skyline. Variable cloudiness

The cold water temperature in the water mains is 18^0 C.

n.p.p. portions of hot water to be drained	The average temperature of the i-th portion of heated water is t_i, C	volume of the i-th portion of heated water V_i, l	Conditional to 37 C heating water volume, $V_{усл.}$	Solar radiation and environmental parameters			
				Time, h	Summarise the radiation drop on the perpendicular. Po w/m. W/m^2 , E	wind speed V, m/sec	ambient water temperature t_0 , C
1	52	10	17,89	9^{30}	810	2,0	16
2	51	10	17,36				
3	49	10	16,43				
4	47	10	15,26	12^{30}	890	1,5	21
5	44	10	13,8				
6	42	10	12,7				
7	39	10	11,05	15^{30}	640	1,5	25
8	37	10	10				
9	36	10	10				
10	36	10	10				
11	35	10	10	18^{30}		1,0	24,7

$Y_{фак} = 110$ litres, $Y_{усл.} = 124.5$ litres.

Measuring the daily output of a solar domestic water heater - the shower cone type.

Place of testing - Fergana

Date 10 May 2012

A qualitative characteristic of the skyline. Cloudless, clear

Cold water temperature in the water supply network 18 C

Table No. 5

n.p.p. portions of hot water to be drained	The average temperature of the i-th portion of heated water is t_i, C	volume of the i-th portion of heated water V_i, l	Conditional to 37 C heating water volume, $V_{усл.}$	Solar radiation and environmental parameters			
				Time, h	Summarise the radiation drop on the perpendicular. In. W/m^2 , E	wind speed V, m/sec	ambient water temperature t_0 , C
1	53	10	18,42	9^{30}	950	1,0	21
2	52	10	18,1				
3	52	10	18,1				
4	49	10	16,2	12^{30}	1003	1,0	26
5	48	10	15,78				
6	47	10	15,26				
7	45	10	14,2	15^{30}	830	1,5	28
8	42	10	12,5				
9	40	10	11,5				
10	38	10	10,52				

11	37	10	10,0	18^{30}	730	0,9	25
$Y_{фак} = 110$ litres, $Y_{усл.} = 160$ litres.							

§2.6. Conclusions from field tests of the BSV-CT.

1. A simple and inexpensive design of a domestic conical solar water heater has been developed, created and field-tested and implemented for individual consumers, and their industrial production has been organised.

2. Field testing of the BSV-CT showed the following:

a) BSV-CT in the climatic conditions of the Ferghana Valley from April to October operates efficiently, with at least 110 litres of hot water with a temperature not lower than $+37^0$ C on a clear sunny day when operating with a storage tank-collector

b) the maximum daily capacity of the BSV-CT on a clear sunny day is obtained in June and July, respectively Vfac = 110 l. and Vfac = 110 l. of hot water at a maximum temperature of $66-68^0$ C. The volume of heated water conditionally reduced to $+37^0$ C is respectively Vsl = 223 l and Vsl = 270 l.

c) The average daily output of the BSW-CT on a clear sunny day obtained in April and October respectively was 110 litres of hot water at a maximum temperature of 54^0 C and 48^0 C.

Overall, the results of the development and in-situ testing of the BSV-CT allow the developed design to be recommended for local heating and domestic hot water supply in central and southern parts of the country.

§ 2.7. Calculation of the technical and economic performance of a domestic solar water heater

To determine the technical and economic performance of a domestic solar water heater, it is necessary to estimate the expected performance and its variation over the year.

It is well known that the performance of a domestic solar water heater depends mainly on its operating temperature, which in turn is determined by the level of solar radiation. The thermal performance of the experimental domestic solar water heater as a function of the operating temperature is given above in the test reports (tables).

The specific annual output of a domestic solar water heater is estimated using the formula

$$q_{год} = q_{\tau} \cdot \tau_{год} \, , \, (1)$$

where, q_{τ} - is the hourly specific capacity of the plant at the average flux density of direct and diffusive solar radiation (we take $E_0 = 700\text{-}800$ W/m^2) and at its required operating temperature;

$\tau_{год}$- Probable number of plant operating hours per year (given the possibility of using the plant in the southern and northern regions of Uzbekistan, the measurement limit, $\tau_{год}$ we take 15003000 h).

From the specific annual output of the plant, the specific fuel savings of a domestic solar water heater can be calculated using the formula:

$$m = \frac{q_{год}}{\eta_{к} \cdot 7000 \cdot 10^3}, \, (2)$$

where, m is the specific fuel savings in tce/m^2 per year;

$\eta_{к}= 0.45$ - fuel utilisation factor in decentralised and individual boilers;

7000 is the calorific value of the reference fuel in kcal/kg.

The results of the calculation are shown in Table 1.

It is well known that the economic effect of introducing a new technique is assessed by comparing the present costs of the proposed and existing options.

Table 1

Specific annual output of a domestic solar water heater and specific annual fuel savings

Number of operating hours of the water heater per year	At operating temperatures up to 37° C		At operating temperatures up to 50-60° C	
	Productivity kWh	Fuel economy T r.t.	Productivity kWh	Savings Fuel T r.t.
1500	1200	0,333	800	0,222
2000	1600	0,444	1060	0,294
2500	2000	0,555	1300	0,361
3000	2400	0,666	1600	0,444

Since the possibilities of using a domestic solar water heater are very wide (hot water supply and heating in industrial, agricultural and communal farms, as well as individual consumers, etc.), the assessment of the economic effect by comparing the reduced costs represents an independent task. Therefore, the calculation of the technical and economic efficiency of the use of a solar parabolic plant was carried out by us according to the payback period, when the economic effect of its use is determined only by fuel savings.

The payback period of an installation can be determined using the formula:

$$C_0 = \frac{K_{уд}}{q \cdot З_т}, \quad (3)$$

where, $K_{уд}$ - specific capital costs, $sиmm/m^2$;

q - specific fuel savings per unit of concentrator surface, t c.e./m^2 ;

Z_T - specific closing costs for fuel (we take 300,000sum, per 1 tce at market prices in 2012).

The new capital costs allocated per household solar water heater (specific capital costs) have been determined using current price lists and guidelines.

The results of the calculation of the specific capital costs for solar domestic water heaters of type VSB-KT. OT-ID/11-4-55 are presented in Table 2.

Table 2.

Total capital costs allocated to one VSB-CT **type solar hot water heater.** OT-ID/11-4-55

№	Name of costs	Value, in sum
1	The storage tank-collector: 1 - galvanised sheet iron, 2 m^2 ; 2 - rubber hose, 1 piece; 3 - float, 1 pc; 4 - transparent heat insulation cap, 1 piece; 5 - mixer tap - shower valve, 1 pc; 6 - Finnish black paint, 2 cans	57,000 30,000 3,000 1,000 5,000 12,000 6,000
2	The shower enclosure frame is quadrangular in shape: a) Steel St.3 pipe, 16 m; b) Fabrication costs for a four-cornered cabin frame - wages of production workers and equipment maintenance costs c) Enamel paint, 1 l	65,000 40,000 20,000 5,000
3	Screen, 2 pieces.	12,000
4	Mounting the unit	5,000
5	Unrecorded costs, 5%	10,000
	Manufacturer's profit 15% of production cost	23,000
	Total cost of the water heater:	172,000

As shown in Table 2, the cost of the water heater is 172,000 UZS (at 2012 prices).

The heat output of a water heater of the type VSB-KT. OT-ID/11-4-55 in summer in the sunny day (without a strong wind) is quite comparable with the productivity of a factory flat solar collector with the area of 2 square metres costing not less than 350 euros or a vacuum tube solar heater, consisting of 20 tubes with diameter of 58 mm and length of 1 metre, with a distance between them of about 60 millimetres - which costs not less than 400 euros. If you compare the cost of an engineered water heater such as VSB-KT. OT-ID/11-4-55 with foreign water heaters, the average cost of $400 is 4-5 times cheaper.

The efficiency of a solar water heater of the type VSB-KT. OT-ID/11-4-55 is caused by the fact that in summer the water heater is illuminated by the sun and accordingly heats not at all 8 hours a day as flat solar collectors, and not 10 hours a day as vacuum solar collectors, but 14 hours. In addition, the VSB-CT heater has a smaller heat exchange surface with the air compared to the aforementioned types of solar heaters.

Assuming that with mass industrial production of such water heaters, the specific capital investment can be reduced by a factor of one and a half, we take the prime cost of the water heater to be 150,000 soums for the calculation of the payback period.

The results of the calculation of the payback period of a solar cone-type domestic hot water heater, depending on the operating temperature and the duration of the installation, are shown in Table 3.

Table 3

Payback **period** for a solar water heater type VSB-CT. OT-ID/11-4-55

Number of	Payback period, one year	
operating hours of	At operating temperatures	At operating temperatures up

the water heater per year	up to 37° C	to 50-60⁰ C
1500	1,8	2,7
2000	1,35	2,04
2500	1,08	1,6
3000	0,9	1,35

Thus, the payback period of a plant is mainly determined by the number of operating hours of the plant (i.e. the number of sunny days per year) and its operating temperature, and is about 0.9 and 1.35 years for $_{max}$ =3000 h, T= 37⁰ C and T= 50-60⁰ C respectively.

Figure 2.11. General view of the organised site for the manufacture and assembly of VSB-CT type domestic solar water heaters.
Fig. 2.12. Examples of a process flow for the manufacture and assembly of VSB-CT type domestic solar water heaters. OT-ID/11-4-55.

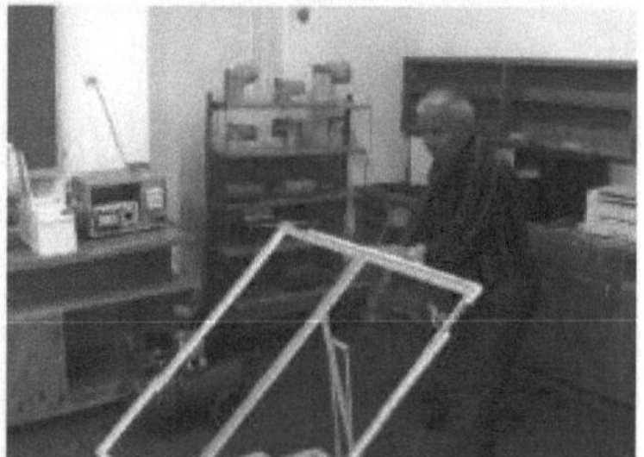

Fig. 2.13. Basic equipment and tools at the manufacturing and assembly site for domestic solar water heaters of the type

VSB-CT

Figure 8.5.Heliopolygon for testing solar installations at the Fergana Polytechnic Institute

Types and technical characteristics of products manufactured on an organised FEP site

1. Solar domestic conical water heater with quadrangular shower enclosure frame.

The main technical features of the water heater:

The storage tank capacity of the solar water heater is -110 litres;
Daily capacity -110 (160 -250) litres;
Water heating temperature - 37-60^0 C
Shower cubicle frame dimensions- 1000x1000x2000 mm.
The weight of the heater without water is 25 kg.

2. Solar domestic tapered water heater with triangular shower enclosure frame.

The main technical features of the water heater:

The storage tank capacity of the solar water heater is -110 litres;
Daily capacity -110 (160 -250) litres;
Water heating temperature - 37-60^0 C
Shower cubicle frame dimensions-1000x1000x2000 mm.
The weight of the heater without water is 15 kg.

3. Solar domestic tapered domestic water heater with triangular base (for domestic hot water supply to individual consumers)

The main technical features of the water heater:

The storage tank capacity of the solar water heater is -110 litres;
Daily output -110 (160-250) litres;
Water heating temperature - 37-60^0 C
The weight of the heater without water is 15 kg.

4. Solar domestic conical water heater for hot water supply to kindergartens, schools, outpatient clinics, backyard and farms)

The main technical features of the water heater:

The storage tank capacity of the solar water heater is 250 litres;
The daily output is 250 (300 - 400) litres;
Water heating temperature - 37-50^0 C
The weight of the heater without water is 20 kg.

5. Solar domestic conical water heater for hot water supply of small and industrial plants.

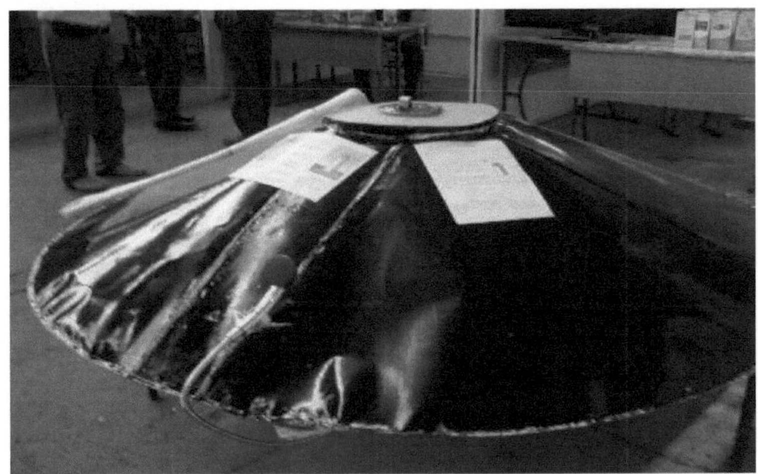

The main technical features of the water heater:

The storage tank capacity of the solar water heater is 528 litres;
Daily capacity -528 (600 -700) litres;
Water heating temperature - 37-50^0 C
The weight of the heater without water is 30 kg.

6. Solar domestic tapered transportable water heater for the small needs of hot water consumers. (laundry washing, dishwashing, hand and face washing).

The main technical features of the water heater:

Solar water heater storage tank capacity -50; 10 litres;

Daily output -50; 10 litres;

Water heating temperature - 37-50^0 C

The weight of the water heater without water is 5; 3 kg.

7. Combined solar plant for the simultaneous generation of electricity and hot water.

Main technical specifications:

The capacity of the storage tank is 250 litres;

Peak electrical power, kW - 1.5 kW per day;

Daily production of hot water, 1 250.

Rechargeable battery 1 piece, 12V, 90 Ah.

CONCLUSION

12.1 At present, a psychological factor prevents the widespread adoption of solar collectors in Uzbekistan. In the near future there will be a rapid filling of this market. Further increases in energy prices in Uzbekistan will force consumers to use new efficient solar heating and hot water systems.

12.2 The advances in the development of solar domestic water heaters are high enough that they are already widely used by individual consumers in American, European and Asian countries. The world leader in production and application is China. In 2007, some 40 million households in China - totalling 150 million people - were using solar water heaters. By 2009, the total area of installed solar water heaters had increased to 140 million m². This is sufficient to supply hot water to around 60 million households. By 2020, 300 million m² of premises in China will be equipped with solar water heaters. The use of water heaters is particularly widespread in Israel, where 95% of flats are equipped with them.

12.3 The largest manufacturer of solar collectors is located in Germany. The solar collector range of TERMOSOLAR (Germany) is the only industrial manufacturer of flat-plate vacuum collectors in the world. THERMO/SOLAR Žiar s.r.o., one of Europe's leading manufacturers of thermal solar collectors, was founded on 1 January 1992. It builds on more than 20 years of experience in the development and production of solar collectors of its founders. The current production capacity is 100,000 m2 of collectors per year.

12.4 Of the modern collectors, German collectors do not require servicing, whereas Chinese collectors need servicing every 2-3 years. However, Asian products are comparable to boiler equipment in terms of cost, and European products are much more expensive. At the same time, German collectors have a service life of 25 years, and they pay for themselves in 10-15 years.

12.5. However, modern water heaters are structurally complex and quite expensive, i.e. economically inefficient for the average consumer. The cost of one m² solar collector for a domestic water heater costs $320-$330. But one collector without basic and additional water equipment will not heat anything.

The calculation of cost of modern foreign systems for heating of water for a family from 4 persons (water consumption in day - 100 litres) - 2 collectors, a boiler, an expansion tank, the control system, accessories will cost $4 000-4 500.

12.6 Widespread use in Uzbekistan of solar water heaters requires more engineering and economic research, the development of design and practical developments and the organisation of their industrial production at affordable prices for individual consumers.

12.7 A simple and inexpensive design of a domestic conical solar water heater has been developed, built and field-tested and introduced for individual consumers, and industrial production has been organised.

12.8. Field testing of the BSV-CT showed the following

a) BSV-CT in the climatic conditions of the Ferghana Valley from April to October operates efficiently, not less than 110 litres of hot water with a temperature not lower than $+37^0$ C on a clear sunny day when operating with a storage tank-collector

b) the maximum daily output of the BSV-CT on a clear sunny day is obtained in June and July, respectively Vfac = 110 l. and Vfac = 110 l. of hot water at a maximum temperature of $66-68^0$ C. The volume of heated water conditionally reduced to $+37^0$ C is respectively Vsl = 223 l and Vsl = 270 l.

c) The average daily output of the BSW-CT on a clear sunny day obtained in April and October respectively was 110 litres of hot water at a maximum temperature of 54^0 C and 48^0 C.

12.9 The heat output of a type VSB-KT. OT-ID/11-4-55 in summer in a sunny day (without a strong wind) is quite comparable with the productivity of factory flat solar collector with area of 2 square meters, costing at least 350 euros, or vacuum tube solar heater, consisting of 20 tubes with diameter of 58 mm, length 1 meter, with a distance between them about 60 millimeters - which costs at least 400 euros. If you compare the cost of an engineered water heater such as

VSB-KT. OT-ID/11-4-55 with foreign water heaters, the average cost of $400 is 4-5 times cheaper.

The efficiency of a solar water heater of the type VSB-CT. OT-ID/11-4-55 is caused by the fact that in summer the water heater is illuminated by the sun and accordingly heats not 8 hours a day as flat solar collectors, and not 10 hours a day as vacuum solar collectors, but 14 hours. In addition, the VSB-CT heater has a smaller heat exchange surface with the air compared to the aforementioned types of solar heaters.

12.10. The results of the calculation of the payback period of the solar conical type domestic water heater depending on the operating temperature are for$_{max}$ =3000 h, T= 37^0 C and T= 50-60^0 C about 0.9 and 1.35 years respectively. For the calculation of the payback period, the prime cost of the water heater is assumed to be 150,000 soums.

12.11. At unitary enterprise "Novator Service" of Fergana Polytechnic Institute together with scientific-production enterprise "Izhodkor" a site for production and assembly of household solar water heaters of VSB-KT type is organized. OT-ID/11-4-55.

12.11. Developed and manufactured prototypes of VSB-KT.OT-ID/11-4-55 type solar water heaters have been demonstrated at IV and V Republican fairs of innovative ideas, technologies and projects held in Tashkent city on April 13-15, 2011 and May 2-5, 2012, at several regional exhibitions, as well as used in about 10 enterprises and institutions of Fergana region.

12.12. Overall, the results of the development and in-situ testing of the BSV-CT allow the developed design to be recommended for local heating and domestic hot water supply in central and southern parts of the country.

REFERENCE LIST

1. Zhimerin D G. Energy. Present and future. - Moscow: Znanie, 1978. - 189 c.
2. the use of solar energy in space research / Translated from English -M.: Mir, 1982. - 402 c.
3. Duffy D.A., Beckman W.A. Thermal processes using solar energy / Edited by Malevsky Y.N. - M.: Mir, 1977. - 409 c.
4. Melnikov Y. F. Light engineering materials. - Moscow: High School, 1976. -151 c.
5. Lurasdaine E., Cherng J. C. On heat exchangers used with solar concentrators // Solar energy. - 1976. - Vol. 18, 1. - P. 197-198.
6. Parmol G. S., Seema L. S. Performance and optimization of a cylindrical parabola collector //Solar Energy. - 1976. - Vol. 18, 3. - P. 150-154.
7. Cobble M. N. Theoretical concentrations for solar furnaces //Solar Energy. - 1961. - Vol. 5, 2. - P. 61-72.
8. Sharver W. W., Duffi W.S. Solar thermal electric power systems composition of lin-focus collectors //Solar Energy. -1979. - Vol. 22, 2. - P. 49-61.
9. Laszlo T. Optical high-temperature furnaces. - Moscow: Mir, 1968. - 207 c.
10. Aparisi R.R. Experimental installation for obtaining high temperatures //Use of solar energy. - Moscow: Izd vo RAN, 1957. - C. 151-152.
11. Baum V.A., Aparisi R.R., Teplyakov D.I. On objective assessment of accuracy of optical systems of solar installations // Thermal power engineering; ed. by Baum V.A. - M.: Publishing house of RAS, 1960. - C. 142-148.
12. Zakhidov R. A., Weiner A. A., Umarov G. Y. Theory and calculation of heliotechnical concentrating systems. - Tashkent: Fan, 1977. - 134 c.
13. Ergashev S.F. Development and Research of Solar Parabolocylindrical Unit with a Heat Pipe as a Heat Receiver: Ph.D. in Technical Sciences. - Ashgabat: NGO-Sun, 1984. - 20 c.
14. Weinberg V. B. Optics in installations for the utilisation of solar energy. - M.: OGIZ, 1959. - 226 c.
15. Durdyev X., Davletov A. et al. Selection of Optimal Parabolocylindrical Concentrator and Receivers in the Form of Pipe // Izv. of Academy of Sciences of Turkmenistan. -Ashgabat, 1977. - №5. - C. 32-40.
16. Evans D. H. On the performance of cylindrical parabolic solar concentrators with flat absorbers //Solar Energy. - 1976. -Vol. 19, 1. - P. 379-385.
17. Teplyakov D. I. Energy characteristics of mirror solar installations under operational conditions // Solar energy converters on semiconductors. - Moscow: Nauka, 1968. - C. 160-165.
18. Teplyakov D. I. Transfer and distribution of radiation in solar installations with mirror concentrators // Solar energy converters in semiconductors. - Moscow: Nauka, 1968 -P. 135-159.
19. Lööf G.O., Foster D., Duffy D. Energy balances of a solar parabolic-cylindrical reflector // Proceedings of the Americal Society of Mechanical Engineering,Ser.A: Power machines and installations. -1962. - T. 33, 1. - C. 33-43.
20. Lööf G.O., Duffy D. Creation of solar focusing collectors of optimal design // Proceedings of the Americal Society of Mechanical Engineering. Ser. A.: Energetic machines and installations. - 1963. - T. 74, 3. -C. 74-83.

21.Aparisi R.R., Garf B.A. Utilisation of solar energy. - Moscow: RAS Publishing House, 1958. - 59 c.

22. Conceptual design and analysis of a 100 MWe distributed line focus solar central power plant: Topical Report / US Department of Energy. - 1978, 1979 -203 p.

23.
Grebenshchikov I.V. et al. Illumination of optics. - MOSCOW: OKHIZ, 1948. - 211 c.

24.
Kokhova I.I., Ergashev S.F. Estimation of efficiency of a tubular receiver of solar radiation // Dokl. of International Symposium on Alternative Energy Sources (Moscow, April 19-20, 1982). - M., 1982. - C. 115-124.

25. Gee R., Gaul H. W., Kearney D., Roalle A. Long-term average performance benefits of parabolic through improvements //SERJ/TR-632 139, March, 1980. - P. 33-38.

26.   Heat
pipes /Translated from English and German. - Moscow: Mir, 1972. - 419 c.

27.
Voronin V. G., Vasiliev A.G. et al. Low Temperature Heat Tubes for Aircraft. - Moscow: Mashinostroenie (Machine Engineering), 1975. - 200 c.

28.   Dan
P., Ray D. Heat pipes /Translated from English - M.: Energy, 1979. - 272 c.

29.
Ivanovsky M. N., Sorokin V. P., Yagodkin I. V. Physical properties of heat pipes. -M.: Atomizdat, 1978. - 255 c.

30. Ivanovsky M. N. et al. Technological bases of heat pipes. - Moscow: Atomizdat, 1980. - 160 c.

31. Vasiliev L.L. et al. Low Temperature Heat Tubes. - Minsk: Nauka i Tekhnika, 1976. - 136 c.

32. Kreeb N., Sohaber K. Heat pipe receiver for a 20 MW solar tower //AIAA Pap. - 1980. - №1507. - P. 1-6.

33. Binert W. Heat pipes for solar energy collectors //Proc. of the 1 st Heat Pipe Conf., Paper 12-1. - Stutgart, 1973. - P. 511.

34. W. B. Heat pipes applied to flat plate solar collectors //Proc. Worshop on Solar collectors for Heating and Cooling of Buildings, 21-23 November, 1974, NSF-RANN-75-019, May, 1975. - P. 620.

35. Ortabasit U., Fehner F. R. Cust mirror-heat pipe evacuated tubulat solar thermal collector //Solar Energy. - 1980. - Vol. 24, 1.- P. 477-489.

36. Feldman K.T., Noreen D.L. Design of a heat pipe absorber for a concen-trating solar water heater //AJAA Pap. - 1980. - №1506. -P. 1-8.

37. Swet C. J. Helitropic thermal generators //The 8th Intersos. Energy Convers. Eng. Conf. Paper no 739128. - New York, 1973. - P. 348-352.

38. Swet C. A. An universal solar kitchen //Proceedings of the 7th IECEC. Paper No729112, American Chemical Society. - Washington, 1972. -P. 720.

39. Kokhova I. I., Ergashev S. F., Sasin V. Y., Borodkin A. A. Heat pipe for solar parabolocylindrical installation // j. Geliotekhnika. - Tashkent, 1982. - №4. -C. 21-26.

40. Sasin V. Я. Heat Transfer Intensity in Test Part of Heat Tubes (in Russian) // MPEI Proceedings. - Issue. 198. - MPEI Publishing House, 1974. - C. 73-79.

41.
Solar oven comes to the kitchen //Macanix illustrated. - 1976. - №7. -P.35.

42.

Muradov D.M. Solar household refrigerator of twenty-four-hour type //Use of solar energy in national economy. - Moscow: Nauka, 1965. - C. 89-91.

43. Solar Industrial Process Heat Conference Proceedings. Oct. 1-Nov. 2. 1979. - Location San Fransisco Bay Area. - San Fransisco, 1979. - P.145-147.

44. Advanced Solar Energy Technology// Newsletter. - 1981. - Vol. 7, 11. - P. 361-366.

45. Bridlik P.M. Testing and calculation of solar desalination installations // Use of solar energy / Edited by Baum B.A. - M.: Publishing house of RAS, 1957. -C. 138-149.

46. Cirardier J. P., Masson N. Use of solar energy in various branches of industry and agriculture abroad // Collection of scientific works of KazNIITI. - Alma-Ata, 1973. - C. 31-52.

47. Aladyev I.T., Kabakov V.I., Kokhova I.I., Ergashev S.F. Test results of solar water-lifting installation with parabolocylindrical collector and jet pump-injector // Proceedings of Dag. ENIN named after G. M. Krzhizhanovsky. - Makhachkala, 1987. - C.135-139.

48. Tarnizhevsky B.V., Kokhova I.I., Ergashev S.F. et al. Tests of power module of solar parabolocylindrical installation //J. Geliotekhnika. - Tashkent, 1982. - №5. - C. 19-24.

49. Kearney D., Daffie J. Bright future for Calif ornian solar plants //Modern Power Systems. - 1988. -Vol.8. -No.7. - P. 31-37.

50. Austrian 10 kWe solar power plant: A Project of the Federal Ministry for science and research/ Published by the Federal Press Service. -Vienna, 1977. - P. 26.

51. Opportunities of using new energy sources // Mater. I International Energy Symposium with Japan. - Irkutsk, 1980. - 170 c.

52. Markman M.A. et al. Experimental study of a parabolocylindrical solar concentrator with a tubular heat sink // J. Heliotekhnika, - Tashkent, 1980. - №6. - C. 66-68.

53. Ramsey J.W., Gupta V.P., Knowles G.R. Experimental evaluation of a cylindrical parabolic solar collector //Transactions of the ASME. - per. S. May , NE. - 1977. - P. 163-168.

54. Kolos Y.G. Investigation of thermal characteristics of parabolocylindrical solar units at different temperatures and pressures of water in the boiler // Proceedings of IV Conference of Young Scientists of RAS . - M.: ENIN, 1957. - Vol.1. - C. 206-207.

55. Edenburn M, W. Performance analysis of a cylindrical parabolic focusing collector and comparison with experimental results // Solar Energy. - 1976. - Vol. 18. - P. 437-444.

56. Barra O., Conti M., Correra L., Visentin R. Thermal regimes in a promary fluid heated by solar energy in a linear collector //JZ nuovo cimento. - 1978. - Vol. 1, 2. - P. 167-184.

57. Huang J., Wund J.Y., Nich S. Thermal analysis of black liquid cylindrical parabolic collector //Solar Energy. - 1979. - Vol. 22. -P. 221-224.

58. Mori Y., Huikata K., Himeno N., Nakayama. Fundamental research on heat transfer performance of solar focusing and tracking collector // Solar Energy. - 1977. - Vol. 19. - P. 50-59.

59. Mikheev M.A. Fundamentals of Heat Transfer. - MOSCOW: OHIO STATE UNIVERSITY OF ECONOMICS, 1956. - 388 c.

60. Kokhova I.I., Malevsky Y.N., Tsvetkov A.I. Method of engineering calculation of STEG // J. Heliotechnica. - Tashkent, 1979. - №6. - C. 22-28.

61. Pogarev G. V., Kiselev N. G. Optical quoting problems. - L.: Mashinostroenie, 1989. - 260 c.

62. Teplyakov D.I. About tolerance system for high-temperature solar installations // j. Heliotekhnika. - Tashkent, 1971. - №3. - C. 17-27.

63. Teplyakov D.I. Engineering Applications of Generalized Radiation Concentration Theory in Paraboloidal Installations of SP and SEU Types // Reports of First All-Union Scientific and Technical Conference on Renewable Energy Sources. Helioenergetica. - Vol. 1. - M., 1972. - C. 282-289.

64. Kalachev P.R. Errors of reflecting surfaces of parabolic mirrors//PhIAN Proceedings. - M., 1970. - VOL. 28, PP. 52-89.

65. Abuev I.M., Tarnizhevsky V.V., Kokhova I.I., Grtsgorian L.3., Ergashev S.F. Energy module of solar parabolocylindrical installation // J. Heliotekhnika. - Tashkent, 1986. - №3. - C. 13-16.

66. Teplyakov D.I. Angular defocusing of parabolocylindrical SES modules // J. Heliotekhnika. - Tashkent, 1987. - №4. - C. 17-22.

67. Yakushev A.I. Interchangeability, standardisation and technical measurements. - Moscow: Mashinostroenie, 1974. - 467 c.

68. Nabibulin F.X., Tarnizhevsky B.V. Method for making faceted concentrators of solar radiation // Solar energy concentrators. - L.: Energy, 1972. - C. 27-32.

69. Rodichev B.Y., Tarnizhevsky B.V. Characteristics of concentrator of solar power plant with photoconverters // J. Heliotekhnika. Geliotekhnika. - Tashkent, 1969. - №2. - C. 9-15.

70. Umarov G.Y., Alimov A.K., Asduazizov A. Investigation of parabolocylindrical concentrator of solar energy from molten glass // J. Heliotekhnika. Geliotekhnika. - Tashkent, 1976. - №6. - C. 52-55.

71 Nabibulin F.X. et al. Method of electroplating in fabrication of parabolic concentrators // Solar energy concentrators. - L.: Energia, 1972. - C. 19-23.

72. Markman M.A. et al. On the angular error of the concentrator // J. Heliotechnica. - Tashkent, 1980. - №5. - C. 20-23.

73. Kokhova I.I., Kabakov V.I., Ergashev S.F., Drobyazgina O.S. Test results of solar parabolocylindrical installation //J. Geliotekhnika. - Tashkent, 1991.-¹2. - C. 14-16.

74. Patent GDR No117107, IPC F03G7/02. Parabolic concentrator. Muller H. (Germany). - C.5.

75. Builov V.V. Mock-up of parabolo-cylinder // Technical Information TS 02.001-80 OKB-1 ENIN named after G.M. Krzhizhanovsky. - M.,1980. -11 c.

76. A. c. 3738774 RF, MPC G2443/02. Reflector / Grigoryan L. 3. et al. - Publ. in B.I. 8. 24.08. 1979. (Russia). - C.8.

77. Abuev I.M. et al. Calculation Technique of Parabolocylindrical Reflector Tray. //Technical Abstracts TS 05.001-84, OKV-1 ENIN named after G.M. Krzhizhanovsky. - M., 1984. - 9 c.

78. Fanak T., Sawata S., Tani T. A terrestrial solar thermal electric power system - Development of basic model system // Solar Energy. - 1977. - Vol.19. - P. 335-341.

79. Treadwell G.N. Design considerations for parabolic cylind. collector // Solar Energy Projects Division, 5712 Albuquceque. -New Mexico, 1979. - P. 35-47.

80. Novikov V.V. On testing solar energy concentrators. - L.: Energiya, 1972. - 41 c.

81. Grichles V.A. Methods of quality control of reflecting surfaces of solar energy concentrators //. Heliotekhnika. - Tashkent, 1972. - №4. - C. 4-15.

82. Aparisi R.R., Kolos Y.G. Adjustment of optical systems of high-temperature solar installations // Thermal installations for use of solar radiation. - Moscow: Nauka. - 1966. - C. 57

83. Tarnizhevsky B.V., Kokhova I.I.. Ergashev S.F., Aliev S.N. Energetic characteristics of experimental module of solar parabolocylindrical installation // J. Geliotekhnika. - Tashkent, 1982. - №6. - C. 25-27.

84. Zakhidov et al. Improved method for adjustment of concentrating systems of autonomous solar power plants with thermodynamic converters // J. Geliotekhnika. - Tashkent, 1989. - №1. - C. 27.

85. Novikov V.V., Skripkar L.N. Concentrator alignment with parabolic facets // J. Heliotechnica. Heliotechnica. - Tashkent, 1969. - №1. -C. 21-23.

86. 3akhidov R.A. et al. Automatic methods of adjustment of facet concentrators and heliostats of solar energy //J. Geliotekhnika. - Tashkent, 1982. - №5. - C. 23-26.

87. Aparisi R.R., Kolos Y.G., Teplyakov D.I. Solar furnaces as a tool for high-temperature research // Thermal installations for using solar energy. - Moscow: Nauka, 1966. - C. 33-53.

88. Laszlo T. Optical high-temperature furnaces. - Moscow: Mir, 1968. -207 c.

89. Lopatina G.G. et al. Optical furnaces. - Moscow: Metallurgy, 1989. - 216 c.

90. Lopatina G.G., Spitsyn V.V. Measurement of Radiant Fluxes in Solar Energy Simulators // J. Heliotekhnika. - Tashkent, 1975. - №3-4. - C. 75-80.

91. Afyan V.V., Shakhparovyan V.V. Photoelectric sensor of concentrated solar radiation density //. Heliotekhnika. - Tashkent, 1975. -№3-4. -C. 139-143.

92. Bazarov B.A., Kapelyushnikova V.M. Experimental setup for measuring spatial and energy characteristics of solar concentrators // j. Heliotekhnika. - Tashkent, 1976. - №6. -C. 30-34.

93. Rauschenbach G. Handbook for Solar Panel Design. - Moscow: Energoatomizdat. 1983. - 351 c.

94. Shcheklein A.V., Recant N.V. Optical research in solar engineering // Solar energy converters on semiconductors. -M.: Nauka, 1968. - C. 184-190.

95. Teplyakov D.I. Investigation of radiation properties of materials on mirror solar installations // Thermal installations for using solar radiation. - Moscow: Nauka, 1966. - C 12-17.

96. Ge Xinshi. Investigation of characteristics of black selective absorbing surfaces // Thermal Power Engineering. - M., 1961. - Issue. 3. - C. 31-37.

97. Mikheev M.A. Methods for Determination of Radiation Coefficient of Solids // Jurn. of Technical Physics. - 1933. - Vol. 3. - Issue. 5. - C. 35-56.

98. Muchnik G.F. et al. Modified non-stationary method for determining the absorptivity of materials // J. Heliotechnica, - Tashkent, 1968. - №5. - C. 16-20.

99. Mavashev Yu.3 et al. Helio-Equipment for Investigation of Radiation Properties of Materials in Wide Temperature Range // J. Heliotekhnika, - Tashkent, 1978. -№6. - C. 33-37.

100. Demidov S.A. et al. A simple portable device for measuring the emissivity of solids at room temperature // J. Heliotechn. Heliotechnica. - Tashkent, 1971. - №6. - C. 37-40.

101. Demidov S.A. et al. Thermoradiometer for measuring the degree of blackness of materials and coatings, calibration and verification of linear scale // Solar Energy Installations. - Moscow: ENIN Publisher, 1974. - C. 131-

102. Anatychuk L. I. Thermocouples and thermoelectric devices. -Kiev: Naukova Dumka, 1979. - 768 c.

103. Belyaev Yu.M. Photoelectric semiconductor actinometric devices // j. Heliotekhnika. - Tashkent, 1990. - №1. - C. 72-75.

104. Phillipov V.L., Popov A.P. Device for providing actinometric measurements // j. Heliotekhnika. - Tashkent, 1974. - №1. - C. 52-54.

105. Osipova V.A. Experimental study of heat exchange processes. - Moscow: Energy, - 1979. - 320 c.

106. Preobrazhensky V. P. P. Thermotechnical Measurements and Instruments. - Moscow: Energia, 1978. - 704 c.

107. Klushin N.P. Laboratory practical work on heat engineering measurements. - Moscow: Higher School, 1970. - 271 c.

108. Mu11er H., Vogel W. Vuk. Bd. 9. Heft 7. -Dusseldorf, 1954. -153 p.

109. Weinberg V.V. Optics in installations for the utilisation of solar energy. - M.: OGIZ, 1959. - 225 c.

110. Abbot C. Obtaining energy by means of solar installations with collecting mirrors // Studies on the use of solar energy. - M.: ML, 1957.-P. 100-106:

111. Garf V.A., Huntzaria R.K. Parabolocylindrical installation with productivity of 40 litres of boiling water per hour // Utilisation of solar energy. - Moscow: Izd vo RAN. 1957. -C. 100-106.

112. Swet C.J. An universal solar kitchen // Proceedings of the 7th JECEC, Paper No. 729112, American Chemical Society. - Washington, 1972. -P. 720.

113. Swet C.J. Heliotropic thermal generators //8th Intersos. Energy Conversion Egg. Conf., Paper #739128. -New York, 1973. - P. 348-352.

114. The solway HPS-200 solar water heater solwa energy copp. 30-942 s. w. Mapine arrive. - Vancouver, 1985. -Vol.6. -P. 572-580.

115. Kusda T. Solar water heating in Japan - Solar Heating and Cooling for Buildings Workshop. - Washington, 1973. - P. 141-144.

116. Proceedings of Informelektro. Issue "Problems of Developing the Market for Solar Power Installations in Capitalist Countries". - M., 1979. - 20 c.

117. Stephen I., Sargent, Barbaro, Gleen H. et a 1. Solar industrial process heat //Environmental Ecology and Technology. - 1980. - Vol.16. - #5. - P. 518-522.

118. Brown K. C., Hooker D. W., Rube A. and t. a 1. End-use matching for solar industrial process heat. Final Report of SERJ. - Japan, 1980. - P. 48-52.

119. Industrial process heat // Solar Age. -1979. -Vol. 4, 3. -P. 19-21.

120. Vindum J., Bents K. Solar energy for industrial process hot water // Agricultural Engineering. - 1977. -Vol. 7. - P. 37-40.

121. Solar Industrial Process Heat Conference Proceedings.Sponsored by the Solar Energy Research Institute for the US Department of Energy. - San Francisco, 1979. -360 p.

122. Advanced Solar Energy Technology. - Newsletter, 1981. -Vol. 7. -№11. -P. 361-366.

123. Solar collectors with cylindrical parabolic mirror// JLBM/NBO. April 6, 1979. - P. 6.

124. Deiser J. Leistungsfahiges ddc-system optimiert den wirkungsgrad ein grosssolaranlage // Linstallatore italieno. - Milan, 1983. -Vol. 34, 2. - P. 240-245.

125. Sequia H., Yoneyama H., Nameda N. Solar total energy system. -UDC 621.311. 001.5:662.997, 1980. - 256 p.

126. Solar energy warms cleaner baths // Industrial Finishing. -1979. - №7. - P. 55,

127. Umarov G.Y. et al. Paraboloid and parabolocylindrical concentrators based on mirror plates of unpolished glass // Reports of All-Union Conference on the Use of Solar Energy. - Erevan, 17-21 June 1969. - M.: VNIIT, 1969. -C. 215-226.

128. Umarov G.Y. et al. Presowing irradiation of cotton seeds by pulse concentrated solar light // J. Heliotekhnika. - Tashkent, 1976, - №5. - C. 47-49.

129. Solar Water Lifting and Transport and Heat Supply Plants // Unconventional Water Lifting and Heat Supply Plants (Prospectus). - Moscow: Vneshtorgizdat, 1989. - 11 c.

130. Tarnizhevsky B.V. et al. To create in experimental manufacture and to test a solar installation for heat supply and water rise on the basis of a focusing collector and a jet pump injector: Report №31/ ENIN, 1986, State Reg.: 01870010360; Inv. №028700180096. - M., 1986. - 49 c.

131. Small solar power system //Acurex Corporation, Alternate energy Division 485 Clyde Ave, Mt View. CA 94042. - Printed in USA. 5/79. -P. 237-245.

132. Solar collectors with cylindrical parabolic mirror //JLBM/MBO. - April, 6, 1979. -P. 6-10.

133. Italian Technical Papers //Italian-Soviet Symposium about the energy renewable sources. -Moscow, April 18-25, 1982. -P. 110-115.

134. Collector Belgo Instruments. - Desquinbei, Antwerpen, 1986. - P. 37-45.

135. Mukhitdinov M.M., Ergashev S.F., Isakulov J. Design features and method of calculation of heat-receiver of solar parabolic tube. //FarPI ilmium-technic zh. 1998. №2. C.70-74.

136. Ergashev S.F. Automated system of measurements, registration and processing of results of energy characteristics of solar systems and environment. Proceedings of International Conference on Photoelectric and Optical Phenomena in Semiconductor Structures (Fergana, 2-3 October 2006). - Fergana, 2006. - C.122-125.

137. Vasiliev V.A., Tarnizhevsky B.V. Calculated technical and economic characteristics of solar combined photothermodynamic power plants. www. solarenergy. ru. - 2004. - 12 c.

138. Ogrebkov D.S. Prospects for the development of solar energy// Russian Chemical Journal. - Vol. XLI. - 1997. -№ 6. - C.43-49.

139. Tarnizhevsky B.V. Assessment of efficiency of solar heat supply application in Russia// Thermal Power Engineering. - 1996, -№5. - C. 57-65.

140. Andreev V.M., Griliches V.A., Rumyantsev V.D., Schwartz M.Z. Photovoltaic modules with radiation concentrators for solar power plants. www.transgasindustry.com. - 2004. - 8 c.

141. Svidersky M.F., Kvaraiheli Y.K. On Development of Work on Solar Energy in the Ministry of Atomic Energy of the Russian Federation. www.transgasindustry.com. - 2004. - 10 c.

142. Volkov E.P., Trusov V.P., Fedorov V.A., Brusnitsyn N.A., Kozlov B.M. Increasing the efficiency of heat conversion technology from unconventional energy sources and converters for them. www.transgasindustry.com. - 2004. - 8 c.

143. Ergashev S.F. Calculation of optimum operation modes of solar water-lifting installation with parabolic collector of solar energy and jet pump injector. //FarPI ilmium-technic zh. - 2005. - №3. - C. 68-72.

Printed by Books on Demand GmbH, Norderstedt / Germany